Democratic Decision-making in the EU

Technocracy in disguise?

This book examines the democratic legitimacy of the European Union (EU) and evaluates the democratic credentials of the EU's main decision-making procedure. It finds that though there is potential for democratic decision-making in the EU, the actual process is dominated by technocrats and secret meetings.

The book assesses and discusses the conditions for democratic input in decision-making with five empirical chapters each addressing the ordinary legislative procedure from different dimensions: democratic deliberative forums, inclusion, openness, power neutralising mechanisms and decision-making capacity. The analytical framework provides for an in-depth assessment of the ordinary legislative procedure's potential democratic qualities and examines whether it fulfils democratic criteria, how the procedure works in practice and whether it has the necessary democratic clout. The author provides both a theoretical discussion and an empirical assessment of what role the principle of democracy could play in the EU.

Filling a gap in EU legislative studies and contributing to the debate on the European democratic deficit, *Democratic Decision-making in the EU: Technocracy in disguise?* will be of interest to students and scholars of European Union politics, legislative studies and deliberative democracy.

Anne Elizabeth Stie is postdoctoral fellow at University of Agder, Norway. She has previously worked at and is still affilliated with ARENA, Centre for European Studies, University of Oslo, Norway.

Routledge studies on democratising Europe

Edited by Erik Oddvar Eriksen and John Erik Fossum
ARENA, University of Oslo

Routledge Studies on Democratising Europe focuses on the prospects for a citizens' Europe by analysing the kind of order that is emerging in Europe. The books in the series take stock of the EU as an entity that has progressed beyond intergovernmentalism and consider how to account for this process and what makes it democratic. The emphasis is on citizenship, constitution-making, public sphere, enlargement, common foreign and security policy, and Europe society.

1 **Developing a Constitution for Europe**
Edited by Erik Oddvar Eriksen, John Erik Fossum and Agustín José Menéndez

2 **Making the European Polity: Reflexive Integration in the EU**
Edited by Erik Oddvar Eriksen

3 **Questioning EU Enlargement**
Edited by Helen Sjursen

4 **The European Union and the Public Sphere: A Communicative Space in the Making?**
Edited by John Erik Fossum and Philip Schlesinger

5 **Law, Democracy and Solidarity in a Post-national Union**
The unsettled political order of Europe
Edited by Erik Oddvar Eriksen, Christian Joerges and Florian Rödl

6 **The New Politics of European Civil Society**
Edited by Ulrike Liebert and Hans-Jörg Trenz

7 **Rethinking Democracy and the European Union**
Edited by Erik Oddvar Eriksen and John Erik Fossum

8 **The Politicization of Europe**
Contesting the constitution in the mass media
Edited by Paul Statham and Hans-Jörg Trenz

9 **Democratic Decision-making in the EU**
Technocracy in disguise?
Anne Elizabeth Stie

Democratic Decision-making in the EU

Technocracy in disguise?

Anne Elizabeth Stie

Routledge
Taylor & Francis Group
LONDON AND NEW YORK

First published 2013
by Routledge
2 Park Square, Milton Park, Abingdon, Oxon OX14 4RN

Simultaneously published in the USA and Canada
by Routledge
711 Third Avenue, New York, NY 10017

Routledge is an imprint of the Taylor & Francis Group, an informa business

British Library Cataloguing in Publication Data
A catalogue record for this book is available from the British Library

Library of Congress Cataloging-in-Publication Data
Stie, Anne Elizabeth.
Democratic decision-making in the EU : technocracy in disguise? / Anne Elizabeth Stie.
p. cm. – (Routledge studies on democratising Europe ; 9)
Includes bibliographical references and index.
1. European Union–Decision making. 2. European Union countries–Politics and government. 3. Technocracy. I. Title.
JN30.S774 2012
341.242'2–dc23 2012025902

ISBN: 978-0-415-52575-6 (hbk)
ISBN: 978-0-203-07642-2 (ebk)

Typeset in Times New Roman
by Wearset Ltd, Boldon, Tyne and Wear

MIX
Paper from responsible sources
FSC® C004839

Printed and bound in Great Britain by
TJ International Ltd, Padstow, Cornwall

Contents

Preface

The former Secretary General of the United Nations, Dag Hammarskjöld, supposedly once said that 'Only he deserves power who every day justifies it.' This quote may sum up the essence of the normative argument defended in this book. It goes to the heart of what democracy should be about, namely a publicly, open and justificatory practice. This book defends not only a *deliberative*, but also a *procedural* approach to democratic legitimacy. The main argument is therefore that a legitimate policy outcome is only partly justified by the quality of its content, it is just as important *how* it was reached. For decision-making processes to be democratically legitimate there should, therefore, be a daily practice of reason-giving and public justification. For this to happen, the institutional and procedural arrangements must be conducive to democratic deliberation. A major task of this book is to specify how we can approximate deliberative principles into mechanisms that *most likely* will maximise the conditions for this happen. Many will, nonetheless, call deliberative democracy utopian or unrealistic. My aim is to show that this is not the case and simultaneously admit that deliberative democracy is hard to achieve and perhaps more fragile than many other models of democracy. However, the difficulties we encounter in bringing it about, does not give us reasons to render the enterprise futile. It rather urges us to investigate the conditions under which deliberative democracy is possible and thoroughly discuss the potential normative trade-offs that can and cannot be made. My hope is that this book can be a small step in this direction.

This book is first and foremost the result of many and highly formative years at ARENA – Centre for European Studies at the University of Oslo. As a research assistant, master student, PhD fellow and later researcher, ARENA has offered direct access to a first class international research environment. In this regard, I have been fortunate to work with Erik Oddvar Eriksen and John Erik Fossum. I wish to thank them both for comments, advice and support. I also wish to express my gratitude to good colleagues and friends at ARENA for collectively creating and maintaining a stimulating and challenging work place. The book has benefited greatly from fruitful discussions, seminars, workshops and conferences in various formats, especially through the collaborative, EU-funded research projects CIDEL and RECON. I am grateful to the many people who have read and commented on the manuscript, in particular, Marianne

Riddervold, Daniel Gaus, Guri Rosén, Helene Sjursen, Chris Lord, Kari M. Osland, Johan P. Olsen, Agustín J. Menéndez, Cathrine Holst and Maria Martens. The present book is based on my PhD thesis and I would, in addition, like to thank Deirdre Curtin and Ulrike Liebert for challenging comments and suggestions both during and after the defence. I am also grateful for comments on the manuscript by the three anonymous reviewers as well as swift assistance from my editor at Routledge, Alexander Quayle. Last, but not least, life is luckily more than research. I am grateful to family and friends for love, support and invaluable distractions.

Abbreviations

ALDE	Alliance for Liberals and Democrats for Europe
CFI	Court of First Instance
Coreper	Committee of permanent representatives
CSO	Civil society organisation
CULT	EP Committee on Culture and Education
DG(s)	Directorate General(s)
ECJ	European Court of Justice
ECON	EP Committee on Economic and Monetary Affairs
EFA	European Free Alliance (cf. The Greens/European Free Alliance)
EMPL	EP Committee on Employment and Social Affairs
ENVI	EP Committee on Environment, Public Health and Food Safety
EO	European Ombudsman
EP	European Parliament
EPP	Group of the European People's Party
EUR-Lex	The online portal to European law
GUE/NGL	Confederal Group of the European United Left/Nordic Green Left
IMCO	EP Committee on Internal Market and Consumer Protection
ITRE	EP Committee on Industry, Research and Energy
JURI	EP Committee on Legal Affairs
LIBE	EP Committee on Civil Liberties, Justice and Home Affairs
NATO	North Atlantic Treaty Organisation
NGO	Non-governmental organisation
OECD	Organisation for Economic Co-operation and Development
OEIL	European Parliament Legislative Observatory
PES	The socialist Group in the European Parliament
Prelex	The Commission's monitoring service of the decision-making process between the institutions
QMV	Qualified Majority Vote
S&D	Group of the Progressive Alliance of Socialists and Democrats in the European Parliament
SEA	Single European Act
TEC	Treaty establishing the European Community

TEU	Treaty establishing the European Union
TFEU	Treaty on the functioning of the European Union
TRAN	EP committee on Transport and Tourism
UN	United Nations
WTO	World Trade Organisation

1 Introduction

One of the main arguments in favour of the Lisbon Treaty was, according to its defenders, that the European Union (EU) would become more democratic. The reason for this stems from the fact that, with Lisbon, the European Parliament (EP) has now gained more influence over EU legislation since the legal areas falling under the co-decision procedure have doubled from around 40 to over 80. This means that the majority of the Union's *legislative* competences[1] need the support of both the EP and the Council. Under the co-decision provisions, the EP is not an appendix to the Council, but an equal partner equipped with the possibility to both amend Commission proposals and to veto them. Consequently, that the EP is now included in EU legislation on a par with the Council in the *vast majority* of EU policy areas, is in itself proof of an unequivocal democratising element, according to the defenders. It may also be argued that with Lisbon this 'fact' is even further underlined by the renaming of the co-decision procedure as the *ordinary legislative procedure* (OLP)[2] signalling and recognising that the normal state of affairs in the EU is that binding legal acts require the support of popularly elected bodies – the EP and the Council (even if the latter only has an indirect popular foundation).

Surely, both with the OLP's inception in the Maastricht Treaty and its subsequent extensions with earlier treaty revisions have the empowerment of the EP has been underlined as a means to strengthen the EU's democratic legitimacy: 'the goal was to enhance the input legitimacy of the EU by giving the only institution whose members are directly elected more powers to shape EU policies' (Häge and Kaeding, 2007: 342; see also Farrell and Héritier, 2003b: 6). However, even if this was the intention, is the inclusion of the EP automatically democratising EU decision-making? Does the EU legislative process suffer from a democratic deficit or is the inclusion of the EP in itself enough? The aim of this book is to answer this question through systematically assessing the democratic quality of the OLP from a deliberative democratic perspective, or more precisely, to answer the following research question: *Are the institutional and procedural qualities regulating the ordinary legislative procedure organised in such a way that the process of co-decision-making warrants the presumption of acceptable and thus democratically legitimate decisions?*

The OLP and the European democratic deficit

Over the years, the OLP has received quite a lot of attention not only within the EU itself, but also from the research community. However, the procedure's democratic credentials have not been the main target and have largely been neglected. Instead, the majority of studies on the OLP are focused on the inter-institutional balance of power (first and foremost) from a game theoretical perspective and often with a special emphasis on the EP's position.[3] Other studies use the development of the OLP as an empirical case to explain institutional change from a rational institutionalist approach (Farrell and Héritier, 2003a, 2004, 2007; Héritier and Reh, 2011; Héritier, 2012; Reh *et al.*, forthcoming 2013). Here, we also find studies focusing on how changes in inter-institutional rules affect intra-institutional arrangements and how the external and internal change processes are connected, including knowledge on how the institutional power balance between the legislative actors – the Council, the EP and the Commission – pattern and change (for instance with treaty revisions or between different decision-making procedures).[4] In addition, there are studies that have a more qualitative approach and often include EU practitioners, usually with the EP as their primary research object.[5] These are important studies because they tell us something about the conditions under which co-decision-making is exercised.

However, EU legislative studies consequently seem to systematically separate the issue of inter-institutional power battles from the question of democracy in the sense that the latter issue is not included in the analytical framework and is hence only treated as an appendix to the main research task. In the OLP literature there are usually only a few comments at the beginning or towards the end of articles where the state of affairs concerning issues of transparency, accountability and legitimacy are mentioned, but they are never the main purpose of the articles.[6] There are – as far as I know – only two studies where the normative implications of the OLP are the main focus.[7] The first appeared in 2003 under the telling title 'The invisible transformation of Codecision. Problems of Democratic Legitimacy', where Farrell and Héritier (2003b) particularly addressed the closed and informal trialogue meetings between a limited number of participants from the Council, the EP and the Commission and how these meetings affect the overall democratic quality of EU decision-making. This is an important study not only because the authors were among the first to thoroughly research the OLP and its specific features, but also because they, with this article, address co-decision-making within a larger normative context of EU decision-making. Another study is the collaborative research project led by Christine Reh on the informal politics of co-decision.[8] This project looks at why, to what extent and for what reasons co-decision actors choose informal and secluded processes rather than following the formal procedural rules. In addition to these questions, the project also deals with the democratic consequences of so-called fast-track legislation, that is, the increasing tendency of adopting OLP acts after first reading. I revisit both studies below.

It is, however, puzzling that the OLP's alleged democratic qualities have not been addressed more thoroughly. After all, the very purpose of introducing the OLP was an attempt to democratise EU legislative processes by including the EP and thus contribute to mending the Union's democratic deficit. With the Lisbon Treaty, the OLP is the most important decision-making procedure in the EU, it has become the default procedure. This is also marked by the fact that the other legislative procedures have been sidelined and now officially figure under the label 'special legislative procedure' (cf. Article 289(2) TFEU). In many ways we could argue that in methodological terms the OLP is the 'most likely' case to study in order to test the democratic quality of the Union since legislative processes resulting in binding legal acts are seen as the prime task or very *raison d'être* of popularly elected bodies in modern democracies. Arguably, whereas the extraordinary events of Intergovernmental Conferences (IGCs) as well as the two conventions leading up to the adoption of the Charter of Fundamental Rights and the Constitutional Treaty respectively are undoubtedly important, the impact of the mundane and everyday processes of piecemeal adoption of ordinary decision-making needs to be addressed more thoroughly. In other words, if the Union fails in the area of ordinary decision-making, which arguably also has a stronger direct effect on the daily lives of Union citizens, the democratic prospects for the EU seem rather bleak.

Moreover, one of the features that truly distinguish the EU from conventional international organisations is the vocal role played by the principle of democracy. Even if unsettled and contested, the problematic state of affairs related to the Union's democratic deficit(s) accompanied by the constant debates about executive dominance and technocracy are, for instance, not in the same way and intensity on top of the agendas in NATO, WTO, OSCE or even the UN. The EU's future is – I would argue – much more vulnerable to expectations and demands on how it can solve its democratic deficits and counter accusations of technocracy and executive dominance than are NATO, WTO, OSCE and the UN (at least for the time being). In facing such demands, the EU resembles democratic states more than conventional international organisations. In short, the normative force of the democratic principle permeates not only the member states, but also the EU – including its legislative procedures. Hence, it is high time to conduct a comprehensive and systematic assessment of the OLP's democratic credentials.

Aim of the book

The aim of this book is to conduct an assessment of OLP where the democratic yardstick is identified, discussed and 'operationalised' in order to make explicit when, and under what kind of circumstances, we can expect that the OLP will produce legitimate and acceptable outcomes. In this regard, I will assess the OLP against the normative standard prescribed by the discourse-theoretical version of deliberative democracy (cf. Habermas, 1998: 107). Hence, if we apply deliberative democracy, when and under what kind of conditions could we expect OLP acts to be democratically recognised? Or put rhetorically and hypothetically, if Union citizens were asked to actually give their consent to OLP acts, what kind of conditions must the OLP ensure in order to make it likely that citizens will find such acts acceptable and thus legitimate constraints on their actions?

In this sense, this book is a contribution to the debate on the alleged democratic deficit, but it differs from other such contributions because it is anchored both in a theoretical as well as an empirical discussion. It provides both a conceptual discussion on the principle of democracy and how this standard can be used when operationalised to the European context *as well as* a systematic empirical assessment of the Union's ordinary legislative procedure. As argued above, not many studies on European democracy provide a comprehensive appraisal where both the theoretical and the empirical components are included in their analyses.[9] Through the analytical framework presented below, this puts me in a position to address the question of what role democracy should play in the EU, both as an idea and as practice, as well as to what extent the current state of affairs qualifies for the democratic deficit category. I do not claim to provide a complete account. As discussed below, the meaning and realisation of the principle of democracy is contested and here I present one version of it – a deliberative democratic account. Hence, even if there are overlapping points, other democratic theories will probably judge the OLP situation differently.

In addition, the book is also a contribution to the 'empirical turn' in deliberative theory. It seeks to operationalise and make more tangible what deliberative democracy could mean not only as idea, but also as practice. As argued below, it is only in the last decade that deliberative democracy has begun to move from almost exclusively being discussed within political theory to also attracting attention from political science. This task of operationalising the concept of deliberative democracy to real life circumstances is still in its infancy and this book is also a contribution to this research endeavour.

Democracy as idea and as practice

This book also helps to abridge what we could call a certain disciplinary division of labour in research on democracy between political theory and political science respectively. Arguably, this could also be part of the reason why EU legislative studies are not primarily concerned with the democratic credentials of the OLP. Basically, this division of labour is related to how democracy is treated in the respective disciplines – as an idea that needs a justification and as practice in real life contexts. We could say that whereas political theory is mainly concerned with how democracy as an idea should be understood and justified, political science has been focused on how democracy as a set of empirical practices unfolds and is exercised in modern democracies.

Very roughly, the debate within *political theory* has for the last two decades been concerned with how democracy as idea is challenged by, and how it can possibly be rescued under, conditions of increasing multiculturalism, globalisation and transnationalisation. Multiculturalism refers to processes inside the traditional nation-state framework and is concerned with how political communities can accommodate difference and heterogeneity. The nation as communal concept has increasingly been challenged as an exclusionary integration mechanism in its demand for homogeneity among its citizenry.[10] Globalisation and transnationalisation processes, on the other hand, are challenges facing the state as the proper unit for problem-solving and democratic decision-making. If societal problems increasingly exceed state borders and can only be solved through states cooperating collectively, can democracy as idea move beyond state borders? In other words, can the idea of democracy be saved when the traditional nation-state framework is challenged? Is democracy possible beyond a nation-state framework, for instance in the EU? With regard to the EU, discussions within political philosophy revolve around whether democracy can be reconciled with and accommodate a system that is multilevel in character, where Union citizens are already democratically constituted in their respective member states, where there is not a functioning public sphere, where there is only a thin common European identity and where decision-making is challenged by technocracy and executive dominance.[11]

In contrast, *political science* is seldom concerned with democracy as idea, but rather seeks to understand how democratic practices unfold and are exercised in and around the political institutions of modern democracies, e.g. electoral and voting behaviour, public sphere, protest movements, civil society organisations/NGOs, lobbying, legislative and decision-making processes, party politics, media coverage, etc. All these topics form important sub-disciplines and most studies focus on one of them. In political science, studies on democratic practices are therefore usually divided or decomposed into bits and pieces, such as how voting behaviour patterns or civil society organisations impact on policy-making. This is also the case in EU studies.

However, when research on democracy as idea and as practice is separated in this way, our understanding of democracy suffers. The problem with political science is that when the analytical task of studying democracy is decomposed into bits and pieces, the essence and idea of democracy as a whole is often lost. What is more, or perhaps as a result of this, empirical studies often take democratic standards as givens or in themselves self-explanatory, focusing solely on empirical data without taking into consideration or discussing that the interpretation of the democratic principle comes in many different versions and that it does matter – also for the empirical analysis – whether one adopts a Schumpeterian or a Habermasian concept of democracy. The principle of democracy is in itself widely accepted and its legitimising force is seldom disputed. It is one of the few principles that have a mobilising effect on people to act collectively to achieve change and recognition, as we have witnessed lately in North Africa. However wide-reaching the agreement on the normative superiority of the principle of democracy as such is, major differences remain about which model is the best one (cf. Lord, 2007b). Hence, how to understand and realise democracy in concrete political settings is neither straightforward nor self-explanatory.

Different perspectives and understandings on what democracy requires and commands in different situations flourish. So when and how do we know that a polity, an outcome, an institution or a procedure is democratic? As indicated above, the answer is 'it depends'. As Bovens *et al.* (2010: 186) argue in their book about accountability:

> It depends on which accountability mechanism, regime, or cluster of regimes is being assessed. It also depends on who is doing the assessing, as there are marked differences among EU scholars in particular as to what constitute the relevant criteria. And it depends on what implicit or explicit benchmarks or counterfactuals are being used in doing the assessing, even among proponents of one and the same criteria set.

'Accountability' is not co-extensive with 'democracy', but the above quote is equally relevant for the latter and 'accountability' could here easily be replaced by 'democracy'. In other words, whether the EU suffers from a democratic deficit (or an accountability deficit) depends on the person who is looking, which empirical case is under investigation, and the democratic perspective(s) chosen. Whereas claims made in studies about the EU suffering from a lack of transparency, accountability, inclusion and deliberation etc. are usually easy to find plausible, the question remains about *how* and *when* they know that a certain process, procedure, arrangement, institution etc. function in a democratically legitimate or illegitimate manner? According to which model of democracy are such conclusions made?

This can be illustrated with the OLP as a case in point, and the aforementioned study by Farrell and Héritier (2003b) who look at the democratic implications of the OLP. Here, the authors use three positions in the debate about democratic legitimacy in the EU as their analytical framework: the first is the federalist position, the second basically amounts to the 'no-demos thesis' position and the third is an in-between position, holding that the current legislative system should in addition be more accountable to, and controlled by, national parliaments.[12] Farrell and Héritier (2003b: 9) call these three positions 'sources of democratic legitimacy', but they never present a democratic yardstick explaining why, when and in what way accountability and transparency are important and necessary criteria for democratic decision-making. They are seemingly just taken as theoretical givens and in themselves self-explanatory. For instance, is accountability understood in the same way from the federalist, the no-demos thesis and national parliament perspectives?[13] How is it possible to determine the putative democratic qualities and/or shortcomings of the OLP when the democratic criteria referred and appealed to – accountability and transparency – are not further accounted for? What are the underlying standards that carry the normative force of statements like these? Without an explicit democratic standard against which OLP decision-making can be assessed, the conclusions about the democratic shortcomings of the OLP remain provisional still awaiting (a more) conclusive and normative endorsement. In other words, what is missing is an argument about 'what role democracy ought to have in the governance of the European Union' (Lord, 2007b: 70).

What is more, many political scientists talk about *legitimacy* and *democratic legitimacy* without making a principled distinction between them in their analytical framework. This does not mean that they are not aware of the difference, but when the research designs do not pick up on the difference the analyses become confusing. For instance problem-solving, efficiency and effectiveness are also important for the overall legitimacy of a political order, but what role do they play in relation to democratic legitimacy? The picture is additionally complicated as different conceptions of democracy underline different solutions as to how these components relate to each other and how the trade-offs between them should be made. Again, the OLP is used to illustrate the case, but this time using the aforementioned study by Reh (2008), which deals with the normative implications of the OLP. The study assesses the increasing tendency to choose the informal and secluded settings of trialogues as a coping strategy under conditions of complexity. The informal politics of the OLP are evaluated against 'three core standards of legitimate decision-making': efficiency, deliberation and accountability. Here, the analysis oscillates between being a general assessment of the normative implications on the one hand, and a seemingly more narrowly defined evaluation of the democratic implications of informal politics under the OLP, on the other. Legitimacy is not co-extensive with democratic legitimacy and it is consequently not clear what precisely is under evaluation. Moreover, the normative standards are not properly operationalised in the sense that the research design does not tell us when the standards are met and when they are not. That is, when do we – analytically – go from accountable to unaccountable? Or when and how does deliberation under the OLP ensure that all arguments are considered and how does efficiency relate to the democratic standard? In sum, the problem is the indeterminacy of what kind of normative assessment the study seeks to provide, that is, an evaluation of whether informal politics under co-decision-making meets the standards of *legitimate* decision-making or *democratic* decision-making.

Another complicating factor is that democratic societies are never made up of democratic institutions and procedures alone. Some institutions and procedures are deliberately non-majoritarian, depoliticised and shielded from the electoral cycles. Others are, unintentionally, not functioning according to the democratic rationale that supposedly underpins them.[14] So, 'how much' democracy should be expected and when are democratic procedures crucial and less crucial?

As illustrated, these challenges and controversies also characterise the debate on the European democratic deficit, but in the EU there is also the additional complicating element related to what kind of order the EU is and consequently what concept of democracy such an order demands.[15] The contested nature of the Union exacerbates the complexity of the debate because there is disagreement about what the 'facts' are about the state of European integration; what these facts tell us about the nature of the Union as a political order and subsequently what these facts command in terms of democratic legitimation. Intergovernmentalists and supranationalists/federalists conceive of the Union differently and consequently disagree about the institutional balance between the central EU bodies as well as the type and scope of competences that should be assigned to the Union and national levels respectively. Whereas intergovernmentalists, for instance, contend that the weight should be put on the member states in the Council, since they represent the democratically organised national demoi, supranationalists/federalists would argue that a supranational decision-making process also needs the authorisation of the directly elected EP. The problem for political scientists is, then, how to situate their empirical findings to the overall meaning of what role democracy plays or should play in the EU. They rarely engage in debates about how empirical findings affect the theoretical assumptions of democratic theory.

Whereas political theory, on its side, is sensitive to the variety of interpretations of the principle of democracy, discussions are too often conceptual and without any anchoring in empirical realities. Often theorisation starts from and is premised on thought experiments such as lifeboat situations, behind a veil of ignorance, ideal speech situations etc. The problem with political theory, then, is that the normative debates are alienated from how democracy as practices, are carried out under real life conditions. Hence, political scientists often question how the insights from political theory should be used. How can highly abstract ideals be relevant under highly non-ideal circumstances such as those offered in the real world? When political theory remains abstract, we also lose insight into what democracy means as actual practice in modern democracies. Surely, the empirical world is constraining on logical and consistent theorisation, but if there is agreement that democracy is not just an abstract idea, but also something that is tangible and possible to realise (even if far from perfect), does not 'ought' then imply 'can'? In other words, should we not expect that political philosophers are bound to reflect on the feasibility of the democratic solutions they defend? To put it a little provocatively, the question is whether normative arguments that are more conducive to empirical analysis should trump or weigh more than normative arguments that cannot be operationalised to empirical conditions without losing their normative bite?

In sum, if political theory and political science take advantage of each other's strengths, the ground for cross-fertilisation and valuable insights increases. This book is a small step in bridging the gap between the two disciplines by showing how it is possible to learn more about democracy both as idea and as empirical practice. As mentioned above, I do not claim to present a complete account of democracy – an account that everyone will agree on, but I do claim that the deliberative perspective on democracy is both relevant and useful in the EU context. Here is my version.

Deliberative democracy

Deliberative democracy is a normative theory based on the idea of:

> a form of government in which free and equal citizens (and their representatives), justify decisions in a process in which they give one another reasons

> that are mutually acceptable and generally accessible, with the aim of reaching conclusions that are binding in the present on all citizens but open to challenge in the future.
>
> (Gutmann and Thompson, 2004: 7; see also Forst, 2001: 346; Chambers, 2003: 308)

Deliberative democracy underlines the importance attached to arrangements where (reversible) policy proposals can be tested through *reason-giving* conducted in *inclusive* and *open* processes *prior* to actual decision-taking. Hence democracy is not only about participation, but also about making sure that the political process ensures quality of decision-making and brings about better results in the rational and epistemic sense. In this way there is, from a deliberative perspective, an internal connection between rationality and deliberation.

However, all representative democracies face the challenge of finding legitimate ways to translate institutionally the democratic requirement of ensuring the inclusion of 'all those affected', given that the citizens never actually participate in public law-making. The deliberative solution is the reliance on satisfactory decision-making procedures:

> [T]he democratic procedure no longer draws its legitimizing force only, indeed not even predominantly, from political participation and the expression of political will, but rather from the general accessibility of a deliberative process whose structure grounds an expectation of rationally acceptable results.
>
> (Habermas, 2001: 110)

The reference to procedure here should be understood in the wider systemic perspective of capturing the constitutional set-up of a democratic order and not just a narrowly defined policy-making procedure producing secondary law (such as the OLP). Rather, the reliance on procedure in the wider sense refers to how the institutionalised decision-making centre is constituted and linked to civil society, the public sphere and to the market (see, Habermas, 1998, 2001, 2006).

Consequently, the deliberative democratic account of this wider perspective on the democratic procedure warrants *decision-making processes* that are sufficiently *open* and *accessible* to the *views* of affected and competent parties. In representative systems the threat is always that those in government become too independent and detached from the control and influence of citizens. This is also a theoretical challenge to which deliberative democracy is subjected. In putting the burden of democratic legitimacy of policy outcomes on the decision-making processes and procedures themselves, the challenge is to ensure that these processes and procedures do not become self-contained or self-sufficient and immune to popular input and scrutiny (Habermas, 1998: 183–4). There must be a link between strong and general publics in order to ensure the subjection of will-formation and decision-making processes in formal decision-making institutions to a third-party audience.[16] The ultimate goal is to ensure that the deliberative processes are not reduced to elite/technocratic deliberation, but are exercised in front of a public where societal voices can respond to and criticise policy proposals through public debate. In other words, a *democratic* decision-making procedure is characterised by *reflexivity*, which means that it ensures institutional and procedural mechanisms spurring internal critique, scrutiny and justification of policy proposals. Consequently, from the perspective of deliberative democracy, then:

> it is not the simple consent that contains the moment of legitimacy but rather a process of accountability in which citizens have objections answered and hear and evaluate the reasons. The approval or disapproval of other people's deliberations reduces democracy to voting.
>
> (Chambers, 2004a: 397)

The EU and deliberative theory

In this book, I assess the OLP from the perspective of deliberative democracy. The main reason for this choice is that several studies have found that deliberation plays an important role for our understanding of European integration. However, the appliance of deliberation and deliberative democracy is often used interchangeably, especially in empirical studies, and this causes confusion.

The difference between Europe and other parts of the world is that the EU member states display a significantly higher degree of willingness and ability to define, solve and cooperate on a strikingly more comprehensive set of issues at the supranational level (see also Curtin, 2009: 3). Moreover, the EU reaches decisions that are complied with, without the traditional state sanctioning mechanisms such as monopoly of violence, (direct) taxation or a common European identity (Eriksen, 2009: Chapter 3). In this regard, many have pointed to the importance of the lengthy and cumbersome processes of reaching consensus among the actors even when a majority vote is possible (Heisenberg, 2005; Lewis, 1998, 2000). Several studies attribute this ability to deliberation and arguing – and not just hard-nosed intergovernmental bargaining – as the important feature explaining why the EU actors are able to reach agreement during various types of decision-making processes.[17] Generally, then, this means that the processes of reaching binding decisions in the EU are marked not only by bargaining and strategic action, but also by the exchange of viewpoints and arguments aiming at justifying and convincing counterparts to change their original position and to reach a common understanding of a given situation.

But these studies do not necessarily talk about the same 'type' of deliberation. For instance, on the one hand, comitology studies are concerned with how networks of experts situated in EU committees are oriented toward intelligent and efficient problem-solving in relation to a cognitive standard and reach decisions through deliberation as a coordinating mechanism, rather than through the mechanism of bargaining (see Cohen and Sabel, 2003; Eriksen, 2003: 208–9). On the other hand, studies concerned with the EU constitution-making process, especially the Convention on the Future of Europe, are also concerned with approximating the mechanism or interaction mode through which the actors potentially reached an agreement on Treaty revision, but the case they describe – constitution-making – is qualitatively different than regulatory and expert committees.[18] What these latter authors want to investigate instead, is whether the process was characterised by less bargaining and more deliberation, and if so, to conclude that constitution-making in the EU was more democratically legitimate or at least more legitimate than the traditional IGC processes. Normatively speaking, then, these two examples do not belong to the same category, even if the actual and observable interaction between the actors can both be labelled deliberative. The point is that these studies refer to different types of deliberative quality. Whereas studies on committees do not claim that the type of deliberation conducted in these settings is democratically legitimate, they are rather, presumably, more inclined to argue that it contributes to 'output' legitimacy in enhancing efficiency and effectiveness and/or that it enhances the epistemic or scientific quality of the outcomes.

In other words, however *deliberative* a decision-making process may be, this does not automatically mean that the outcome is *democratically* legitimised (Eriksen, 2003: 211, 217; see also Nanz and Steffek, 2005: 268–9). Whereas deliberative democracy is a normative theory prescribing norms and ideals for the realisation of democratic government, the deliberative theoretical field is not co-extensive with deliberative democracy. Building on the Habermasian notion of communicative action,[19] deliberation can also simply be used as an explanatory theory of action coordination – and thus function as an alternative explanation to rationalist theories based on bargaining and strategic action.

Deliberative theory obviously displays an analytical power to capture central elements of European integration and has thus become a popular framework among many EU scholars. But despite these theoretical strengths, there is also a tension or confusion with regard to its analytical appliance; especially to empirical conditions. This confusion amounts to the difference between deliberative theory as an action theory, versus deliberative theory as a theory of democracy. Hence, when the purpose is not (only) to *explain* the likelihood of whether a result or agreement was reached through the mechanism of bargaining or deliberation, but rather to *evaluate* the *democratic* qualities of the decision-making process, it is important to be aware of the three different qualities/dimensions of deliberation: (1) As a mode of interaction deliberation is oriented towards reaching a common understanding among the participants about the nature of a given case. Hence it is an alternative interaction mode to bargaining, which is based on strategic action. The focus is on how deliberation has the potential for testing and possibly transforming and changing preferences/positions. (2) As an interaction mode, deliberation is also highlighted as a mechanism conducive to problem-solving, and to enhancing the epistemic quality of decisions. The epistemic quality of deliberation contributes to bringing out the cognitive or expert (scientific or technical) knowledge aspect of a case.[20] (3) Finally and most importantly, deliberation is – given the fulfilment of certain institutional and procedural preconditions – also a conflict-regulating mechanism that is suitable for determining questions of justice and impartiality and consequently providing a procedure for reaching legitimate collective decisions. This level of deliberation is normatively more demanding as the moral content of deliberation can only be redeemed if it ensures the inclusion of all those affected by the decisions (cf. Habermas, 1998: 107; Eriksen, 2005: 16–17).[21] The realisation of the third quality of deliberation points to the importance of institutions and procedures.

This book starts from the assumption that the way decision-making processes are organised and institutionalised can tell us something about the likelihood of whether a procedure is (at all) *able* to generate democratically acceptable results. Focusing on procedural qualities, then, means to look at the conditions and the structure of the decision-making framework. For instance, if decision-makers are not in any shape or form subjected to two widely shared minimum requirements, namely periodic elections (i.e. a procedural mechanism regulating access to decision-making posts) and public debate, the outcomes resulting from the decision-making process can simply not be labelled *democratic* as the two most important conditions for democratic law-making have not been met. Institutional and procedural features matter as they shape the behaviour of actors and consequently also the outcome of interaction – they function as 'rules of the game'. My focus in this book, then, is on how the procedural design and the informal rules that follow from it do, or do not, function as constraints, possibilities and incentive-structures for ensuring and promoting democratic decision-making.

How to approach deliberative democracy empirically?

The purpose of this study is to develop an evaluative scheme for the assessment of the OLP. On a more general level it will also present a scheme that can be applicable to other empirical studies seeking to evaluate the democratic quality of public policy-making procedures. Hence, the aim is to establish a first step in showing that it is possible (and important) to distinguish empirically between the different types of deliberative qualities and thus to facilitate a more nuanced normative account of what deliberation can contribute with in different types of situations; that is, whether it is only likely to enhance problem-solving or whether it also has the potential to enhance the democratic quality of collective decision-making.

Over the last two decades, the theory of deliberative democracy has not only received major philosophical attention,[22] but has, as we have seen, also become an increasingly important perspective on European integration.[23] In addition, there are also other empirical approaches which do not have the EU as their main target, but that deal with the operationalisation of deliberative theory more generally.[24] Whereas the *philosophical* debate on these issues has lasted longer and is thus already sophisticated and advanced, the task of *operationalising* and thus making deliberative theory relevant for research on concrete empirical cases (e.g. European integration) is still under-developed. As argued, this book aims to contribute to the on-going endeavour to operationalise deliberative theory to non-ideal conditions. This is important not only because it helps to test empirically the claims and assumptions of the theory of deliberative democracy, but also because empirical research can help to correct, nuance and further substantiate the kinds of conditions under which deliberative democracy is applicable, viable and desirable.

Deliberation can be a good thing in many situations; it can be useful in the private sphere and in the public sphere, as well as in formally institutionalised decision-making settings. With the different types of deliberation in mind, we must also be careful about taking heed of the context or situation in which deliberative democracy is supposed to be applied. When, where and how deliberative democracy is used is critical for the types of conclusions that can be reached. Thompson (2008: 503) notes that 'Until recently nearly all studies – and much of normative theory – investigated deliberation by ordinary citizens rather than politicians.' Consequently, given that the majority of empirical studies investigating the effect of deliberation are concerned with small groups and do not focus on actual public decision-making, this limits their value 'not only because what is being studied is several steps removed from what deliberative theory is ultimately concerned about,[25] but also because discussion alone is likely to produce different empirical consequences than those of decision-oriented discussion' (Thompson, 2008: 513). In other words, the conclusions and findings resulting from studies of small groups without the power to make binding decisions cannot, and should not, automatically be generalised as valid for institutionalised decision-making settings.

Consequently, it is important to specify the circumstances in what could and should be expected normatively in different situations. There is for instance, a major difference between what can and what should be expected from actors in institutionalised decision-making settings as opposed to actors participating in public sphere experiments such as 'Deliberation Day', 'Deliberative Poll' or other types of citizens' assemblies (Ackerman and Fishkin, 2004; Fishkin, 1999; Fishkin and Luskin, 2005; see also Warren and Pearse, 2008). For participants in the first type of setting, decision-making resulting in binding decisions is involved, that is, will-formation processes that have a strong connection and lead directly to binding decisions. For the latter types of settings the connection between deliberation and decision-making is at best very weak; that is, deliberation functions here merely as opinion-formation because it does not directly affect the setting of norms/goals and the (re-)distribution of interests, i.e. such discussions are non-binding upon the participants. This does not mean that studies of smaller groups or debates in the public sphere are unimportant, they are all relevant for the larger democratic process. The point is that a research design must be carefully devised so that it suits the object of study (Thompson, 2008; Warren, 2007: 286). What, then, should we expect from a decision-making procedure such as the OLP?

The OLP belongs to the category of institutionalised decision-making settings and consequently involves both opinion- and will-formation, as well as decision-making processes. Moreover, the outcomes of OLP processes are EU regulations, directives or decisions, that is, they are binding on the addressees. The first requirement, then, is that the provisions regulating co-decision-making have allocated time and space so that there is (at least) a chance for deliberation to occur prior to actual decision-making. Second, and as we have seen, the goal of deliberative democracy is not only to ensure that there is reason-giving and deliberation, but also that the reasons or deliberation are of a certain quality. However, in order to determine the democratic quality of decisions it is not enough only to assess the validity and consistency of the speech acts or arguments themselves. The democratic quality is first and foremost redeemed when and if it can be ascertained that the context in which these arguments were uttered meet certain procedural standards for democratic decision-making such as inclusion, transparency etc. The quality of decisions is primarily determined and related to the overall quality of the setting or the conditions under which they are presented. In other words, outcomes resulting from deliberation in closed settings with flexible mandates, which do not warrant open accountability and justificatory processes (neither during nor after the meetings), have a normatively inferior status compared to outcomes resulting from settings where policy proposals must withstand critical scrutiny in a process that is either open during or after the meeting. Third, the final requirement is to make deliberation relevant for decision-making, that is, argumentation should have an impact on the outcome.

Deliberative democracy applied to institutional decision-making settings

The aim of this book is to assess how far the OLP institutional arrangements meet the *procedural conditions* for legitimation through democratic deliberation between elected representatives. Hence, in order to avoid confusion it should immediately be underlined that the task is not to assess whether co-decision-making is in fact characterised by deliberation rather than bargaining, or to assess the quality of the putative deliberations. Instead, this book is concerned with evaluating the democratic quality of the conditions under which co-decision-making is exercised and, more precisely, to answer the following research question: *Are the institutional and procedural qualities regulating the ordinary legislative procedure organised in such a way that the process of co-decision-making warrants the presumption of acceptable and thus democratically legitimate decisions?* In other words, is the OLP organised in such a way that it makes democratic deliberation and discussion possible prior to final decision-making? And if so, does the putative deliberation taking place in these debates approximate standards of democratic reason-giving? An important distinction in this regard is the difference between elite/expert and democratic deliberation. To put it bluntly, the latter question pertains to finding out whether discussions are only between experts and bureaucrats or whether they also open up for the viewpoints of affected parties.

The way to determine these differences in the normative quality of deliberation is to pay attention first and foremost to the actors and how – given the procedural formal and informal rules – they (are likely to) behave in different settings. First of all, there is a major distinction between elected and non-elected participants, and it is only the former who are eligible for the normatively highest status as representatives of affected parties. Moreover, it is only they who are – through elections – authorised to make collective decisions on behalf of citizens and who thus possess the key to connect the deliberation that takes place inside institutionalised decision-making settings to the public deliberations in the general publics outside the formal decision-making nexus. However, even if the electoral link is a necessary precondition, it is not sufficient. Elected actors must earn their status as representatives of affected parties on a daily basis by the way they behave. They must justify their positions in clear and intelligent ways, they must conduct their business in openness and they must ensure that all types of views are voiced and heard. In this regard we could think of a rephrasing of Ernest Renan's famous definition of the nation as a 'daily plebiscite' with Dag Hammarskjöld's 'Only he deserves power who every day justifies it.' In other words, the elected decision-makers should think of their work as a daily justificatory and account-giving practice.[26] As we can see, the justificatory duty put on individual elected representatives is tied to the democratic requirement of public debate which again provides a link back to elections as the mechanism that authorises elected representatives (see also Eriksen, 2009: 229).

However, democracy is not only a principle of legitimation, it also warrants an organisational interpretation in order to realise its aspirations. The organisational structure of modern democracies has systems to translate principles and arguments into action. There are rules for law-making, for representation, for argumentation, for how to sanction power abuse etc. and there are resources allocated to make sure that these goals are realised. In other words, to put such a heavy burden solely on the representatives alone without also institutionalising the above principles into formal rules and practices is too demanding. Hence to avoid weakness of will in situations where it may be tempting to revert to closed and unaccountable settings, the institutionalisation of formal and observable rules also provides necessary mechanisms to help ensure that elected participants behave in accordance with what is expected from those who are selected to represent affected parties. Hence the procedural rules should institutionalise mechanisms that ensure inclusion, openness, accountability and equality.

There is a major debate about how to provide empirically researchable indicators that can distinguish between deliberation and bargaining in real-life settings. As the purpose of this book is to assess the procedural qualities of the OLP, this means looking at the *preconditions* for deliberative democracy, that is, whether the procedural circumstances regulating co-decision-making – *in general* – are more or less *likely* to contribute to democratically legitimate decisions. In other words, I do not assess the interaction between the Commission, the EP and the Council in the adoption of *concrete* OLP dossiers in order to determine whether the policy-making processes were in fact characterised more or less by deliberation (or bargaining). Moreover, I do not assess whether the quality of the actual arguments in OLP debates come closer to expert/technical or to political deliberation. This is not the task of this book. Hence, here the debate on how to empirically distinguish between deliberation and bargaining is only indirectly relevant, but it functions as an important background context. In this regard, and as emphasised above, my contribution to this debate on developing indicators that can tell deliberation from bargaining when we see it, is limited to underlining that in order to know what kind of deliberation we potentially face when studying concrete empirical cases we need to investigate more than just the arguments or utterances. We also need to know the structural *conditions* under which they were expressed. In other words, merely investigating speech acts is insufficient for *determining empirically* whether a case was characterised first and foremost by one or the other interaction mode, because only *the way* in which arguments and counter arguments are included and scrutinised, can tell us whether the decision-making result deserves democratic recognition.[27] In sum, the overall structure or context of deliberation and decision-making must be taken into account.

Outline of the book

There are two main parts in the book. The first is devoted to the development of the theoretical framework as well as a background presentation of the OLP (Chapters 2–4). The second part contains the evaluation proper (Chapters 5–10).

There are three chapters in the first part. Chapter 2 introduces the theory of deliberative democracy with a special emphasis on the discourse theoretical version, which is applied in this book. This chapter describes the democratic standard against which the OLP will be evaluated. More precisely, the deliberative approach to democracy is first contrasted to bargaining and aggregative approaches underlining the deliberative democracy as a 'talk-centric' (as opposed to 'vote-centric') theory where accountability is conceived as an ongoing reason-giving practice. Here I also discuss the types of outcomes that can be expected from deliberative democratic processes (from rational consensus to working agreement) and how this can be achieved, that is, the need for an egalitarian law structure. In the second part of Chapter 2, I address how other studies have approximated deliberative democracy and what can be learned from these and potentially applied to the evaluation of the OLP. As we have seen, many empirical approaches do not take proper heed of the difference between the conditions under which deliberation can lead to democratic government and when they merely lead to deliberative governance (that is, ensuring efficiency, problem-solving and/or improving the cognitive-scientific foundation for decisions).

In Chapter 3, the understanding of democratic legitimacy is approximated from being solely a legitimacy principle to also becoming an organisational principle. Here I present an evaluative scheme containing five normative criteria, which are subsequently 'operationalised' to empirical indicators. The criteria are:

1 democratic deliberative meeting places;
2 inclusion of affected and competent parties;

3 openness;
4 neutralisation of asymmetrical power relations;
5 decision-making capacity.

In the last part of this chapter, methodological issues are raised.

Chapter 4 provides background information on the OLP. Here the legal provisions and stages of the procedure are presented as well as a brief historical outline of the major developments and changes to the OLP since its inception in the Maastricht Treaty. Furthermore, it is supplemented with some numbers on how the OLP performs.

The second part of the book holds the evaluation. Following the criteria, there is one chapter for each criterion. Chapter 5, then, addresses the requirement of *democratic deliberative meeting places.* A minimum requirement for such meeting places is that they are composed of a representative selection of elected representatives. Only five OLP settings meet this initial threshold, hence in this chapter the following settings are considered: the EP committees, the EP plenary sessions, the EP conciliation delegation, the ministerial level in the Council and the Conciliation Committee. The question is whether these meeting places are settings where discussion, testing and justification of positions on OLP dossiers are/can be conducted in a publicly accessible manner.

In Chapter 6, I move on to evaluate the conditions for *inclusion of the views of affected and competent parties*. The first indicator evaluates whether the elected institutional actors as collective bodies possess final veto power. The second indicator is more demanding in assessing whether the OLP respects the normative hierarchy between elected and unelected actors on an individual level, that is, if it looks at how the division of labour inside the institutional actors patterns between experts and elected politicians.

The seventh chapter is devoted to the issue of *openness and transparency*. At a minimum, the OLP proposal should be available before the decision-making process is over. This is the first indicator. The second indicator builds on the first and appraises the conditions for access to documents and whether they are stored and easily available in a digital register. The third indicator looks at the openness of debates whereas the fourth deals with the intelligibility of voting.

In Chapter 8, the focus is on mechanisms that can *neutralise asymmetrical power relations*. In this regard, two ex ante mechanisms are considered: intelligibility of procedure and separation of powers. In addition, I look at two ex post mechanisms that can help to neutralise unequal power conditions, namely, the possibility for judicial review and/or recourse to the soft law possibility represented by the European Ombudsman.

In Chapter 9, the fifth and final criterion of *decision-making capacity* is addressed. The issue to be assessed here is whether there are informal networks or structures where OLP acts are actually taken, and consequently whether the formal decision-making framework assigned in Article 294 (TFEU) is by-passed and at worst made redundant. Two potential candidates are appraised. First, historically, legally and actually, the Council is the stronger party in relation to the

EP. It is therefore not unlikely that the Council would in fact also be the locus where decisions were taken and thus reduce the EP to rubber-stamp take-it-or-leave-it deals. Second, throughout the book the informal meetings – so-called 'trialogues' – between the EP, the Council and the Commission are a recurring topic. The importance of these meetings for reaching agreement is continuously underlined and they are now taking place at all stages in parallel with the formal processes. The question to be discussed is therefore whether they have taken over and made the formal processes redundant apart from ensuring formal approval through the vote.

Finally, Chapter 10 holds the concluding remarks summing up the findings and the lessons that can be learned by using the evaluative scheme. Here I first discuss the OLP's democratic qualities and subsequently its democratic deficits. Given the severity of the latter, it is argued that the former can only be perceived as potentials rather than how the OLP is actually practiced. The last part of the chapter is devoted to a discussion on the strengths and weaknesses of the evaluative framework – on what it permits us to conclude and in which direction it can be improved and further developed.

Part I

Theory, approximation and background

2 What is democratic legitimacy?

Empirical enquiry can more effectively influence – and in turn be influenced by – normative theory if both theorists and empiricists proceed with a clearer conception of the elements of deliberation.

(Thompson, 2008: 498)

Introduction

In the previous chapter I briefly introduced the concept of deliberative democracy and how the OLP could be assessed for its potential democratic qualities against five criteria (democratic deliberative meeting places, inclusion, openness, neutralisation of asymmetrical power relations, and decision-making capacity) developed from this theoretical position. The purpose of this chapter is to present in a more systematic manner the deliberative version of democratic legitimacy on which these criteria rest. Over the years, deliberative democracy has become a large research field and it is consequently impossible to do justice to all positions here,[1] but the account of deliberative democracy defended in this book is based on the Habermasian (1998, 2001, 2006) discourse theoretical version and supplemented primarily with Eriksen and Weigård (2003), Eriksen (2003, 2009); Gutmann and Thompson (2002, 2004), and Chambers (2003, 2004a, 2004b, 2005).

The chapter contains two parts. In the first part, I present the deliberative understanding of democratic legitimacy, which, in short, is 'talk-centric', emphasising the institutionalisation of practices of accountability and justification. Here I discuss the types of outcomes that should be expected from democratic processes (from rational consensus to working agreement), whether the deliberative model of democracy is realistic as well as the need for an egalitarian law structure. The second part is focused on how other studies have operationalised deliberative democracy and what can be learned from these and potentially applied to the evaluation of the OLP. The problem with many empirical approaches is that they, as alluded to in the first chapter, do not pay enough attention to the difference between the conditions under which deliberation can lead to democratic government versus merely deliberative governance. Or, put differently, when can deliberation qualify as an expression of democratic legitimacy and when is it only contributing to efficiency, problem-solving and/or improving the cognitive-scientific foundation for decisions, that is, for deliberative governance? In sum, the crux of the matter relates to the type of evaluation standards that are being applied.

Democratic legitimacy from a deliberative perspective

The modern idea of democracy basically amounts to the notions of the self-governing human being and a just society where free and equal individuals give themselves laws that in the next step can be seen as legitimate constraints on their actions. How can this ideal scenario honouring the principle of individual autonomy be translated to real life circumstances? Under conditions of pluralism, where there is no prospect of reaching a consensus on how everyone should live their life, the way to *determine* whether decisions are legitimate cannot merely be about their substance. The chance that everyone will find a policy proposal equally good is unrealistic. Consequently, given that the substance of a proposal cannot function as the sole motivation to support a decision, the deliberative argument is that the legitimacy of decisions must first and foremost rest on the *decision-making procedure itself*.[2]

The assumption is that if an issue has been properly treated in a fair process, it is likely that those who were opposed to the outcome will nevertheless respect it as a legitimate constraint on their behaviour despite the fact that their position was rejected in the final decision-taking moment. The theory of deliberative democracy assumes that if there is a prospect for such a fair process preceding decision-making, the majority vote can be democratically legitimised. Moreover, when the requirement of a prior process is respected, majority votes 'therefore represent only temporary stops in the continuous discussion about what should be done. Such a procedural interpretation of the majority principle makes it consistent with the concept of freedom when not applied to irreversible decisions.' (Eriksen, 2009: 48; cf. also Chambers, 2004a: 397). Without such prior processes, the actual act of voting becomes nothing more than an arithmetic number at a given point in time (Eriksen, 2007a: 92–3). If fair procedures are established they can provide the reason and motivation for minority positions to support majority decisions and can thus compensate for the unrealistic prospect that everyone will come to agreement on the substance of a decision.

The difference between the theory of deliberative democracy and other theories of democracy is consequently locus. Deliberative democracy is 'talk-centric' not 'vote-centric' (Chambers, 2003: 308). To emphasise the process preceding the actual decision-making moment differs significantly from aggregative theories of democracy, where the aim is to find the best mechanism to translate already fixed individual preferences as accurately as possible into a collective decision by ensuring that every citizen counts for one vote, and only one vote, in the final outcome (ibid; Nanz and Steffek, 2005: 370). The argument of the theory of deliberative democracy is that preferences are not given once and for all – they are not impermeable of human interaction, but are rather shaped socially and discovered and influenced by the very process of interacting with others. What is more, even in situations where preferences are not altered through social interaction, the preferences of the majority still need to be justified and legitimised through reason-giving so that the minority can know that their views were set aside for better reasons, instead of just by-passed by a mere act of will (Dewey, 1927; Mill, 1861). Hence, a prior deliberative process is not pointless, but bears the hope and anticipation that divergent positions can be collectively moulded and potentially transformed into a consensus.[3] For this to happen, it is important that the process itself is regulated in such a way that different views, desires and knowledge can dominate the process rather than partial arguments, unjustified positions and insufficient knowledge. The problem with vote-centric theories is the lack of a mechanism that can ensure a just and non-arbitrary way of aggregating individual preferences.[4] Aggregative theories thus not only lack a mechanism to take heed of the conflict regulating, learning and problem-solving potential of conducting a deliberative process prior to actual decision-taking. More importantly, they lack a mechanism to *test* the *quality* of *individual preferences* and consequently also a basis on which political outcomes can be justified and legitimised towards those whose positions did not win the vote, but must nevertheless accept and respect the law as binding on their actions. By testing and 'laundering' preferences through open public debate the presumption is that the remaining positions, which have withstood critical scrutiny, can also be accepted by those who initially were against. Purely self-interested preferences do usually not survive this kind of public and inclusive interrogation unless those advocating them are in an extremely disadvantageous position.[5] In short, deliberative democracy 'moves the heart of democracy away from the vote and into the public sphere and practices of accountability and justification' (Chambers, 2003: 311).[6]

Even if bargaining-driven theories are more tuned in – than are vote-centric theories[7] – on the potential represented by conducting a collective process prior to the decision-making moment, they still revolve – as do vote-centric theories – around a rational choice logic where actors are perceived as utility maximisers, only conceding to others what they deem necessary given their relative bargaining strength. Consequently, what takes place prior to decision-making is not what deliberative theorists would call deliberation or arguing, but bargaining and strategic action and the outcomes of bargaining processes are, by deliberative democracy, defined as *compromises* and not true agreements because the logic inherent in rational choice assumes the alteration of the outcome the moment there is a change in the relative bargaining strength among the actors. Hence, outcomes reached through bargaining are inherently unstable and thus, even from a purely instrumental perspective, there are good reasons to strive for other, more stable types of outcomes. However, and more importantly from a democratic perspective, they are also potentially unfair because in basing their practical decision-making arrangements on a bargaining logic (i.e. that self-interested actors will maximise their self-interest), such arrangements only perpetuate the power inequalities among the actors as the participants are not expected or encouraged to take into consideration the needs and arguments of others (unless it is in their interest to do so) and see themselves as part of a collective process, where an outcome should be acceptable and reasonable for *all* the parties involved (see Miller, 2000). This does not mean that bargaining procedures have no place in a decision-making process, it only means that the entire process cannot be based on a bargaining logic alone.[8] And more importantly, such arrangements must be (discursively) justified in a prior process as only procedurally regulated bargaining arrangements can claim legitimacy.

How can the political process be arranged in order to reduce or compensate for the power inequalities between actors? According to the theory of deliberative democracy, the key is to maximise the likelihood that actors will take their decisions argumentatively, that is, on the basis of giving, receiving and possibly being convinced by arguments and good reasons, rather than the actors first and foremost taking into the equation the size and strength of opponents and thus the threats of being punished if they do not adhere to what stronger actors want. Or alternatively, if they are the stronger party, that they will abstain from using threats and punishments to get their way. In other words, a democratic decision should not be based on extra-argumentative elements, but rather, as far as possible, must be based on good reasons, as this provides a much more stable and legitimate foundation for public policy-making. Moreover, by its very nature deliberation, as a conflict coordinating mechanism, stands a much better chance – than do other mechanisms such as money, power, tradition, religion etc. – of respecting democratic principles such as inclusion, autonomy, freedom and respect for individuals (Warren, 2007: 274).

There is, however, an additional reason why deliberation should be the dominant mode of interaction during decision-making procedures. Deliberative processes not only increase the likelihood of respecting the principle of equal consideration of all viewpoints, such processes are also apt to ensure the epistemic and cognitive quality of decisions. As mentioned, deliberation is also a mechanism for problem-solving, fact-finding and fact-sharing which can enhance the knowledge foundation of decision-making. In this way, deliberation is not only an interaction mode facilitating the distribution of burdens and benefits (as is the case with bargaining), it is also a mechanism to clarify differences and potential misunderstandings, to shed light on facts and elaborate on ethical and moral dilemmas. In sum, deliberative processes can (at least in principle) enhance the likelihood that policy proposals will be put both to an impartiality test as well as to a cognitive test (Eriksen, 2009: 44). But when do we know that decisions have passed these tests, that is, when do we know that decisions have been reached legitimately?

From rational consensus to working agreements

We must first of all identify the normative/democratic standard that must be respected. In discourse theory this amounts to the discourse principle: 'Just those action norms are valid to which all possibly affected persons could agree as participants in rational discourses' (Habermas, 1998: 107). The discourse principle establishes that legitimate decisions are those that can be consented to after withstanding scrutiny and critique by those who are going to be bound by them. In representative democracies it is, however, not the case that all citizens and all affected parties are involved in actual decision-making. Hence the formulation '*could* agree to' may be taken to indicate that the principle can be used as a hypothetical litmus test for assessing the legitimacy of actual decisions (ibid., my emphasis). It should immediately be added that the hypothetical test cannot replace actual public discourse (Eriksen, 2007a). A viable democracy cannot do without public debate where policy proposals are tested and discussed in front of, and involving, the public at large. What is more, the discourse principle implies that legitimate decisions are achieved if certain conditions for discourse are met. When translated to the institutional and procedural level, the discourse principle contains 'four ideal claims on institutions and processes: freedom, rationality, equality and publicity' (Eriksen, 2007a: 95–6).[9] The bottom line is that democratically legitimate decisions have been *validated* through a *particular type of process* which entails that they have been defended, tested and criticised argumentatively in a publicly accessible debate and that minority positions have been included, listened to and taken into account during the course of a collective and inclusive process (Chambers, 2004b). Decisions that are to be labelled democratic have to comply with elements of the four procedural requirements presented above. But how, more concretely, can these ideal claims be approximated to non-ideal conditions?

The normative standard for democratically legitimate decisions is the 'rational consensus', i.e. an outcome the involved actors can support with identical reasons (Habermas, 1998: 339). To evaluate whether an outcome fulfils or falls short of a rational consensus, a decision must be assessed against the criteria of an 'ideal procedure' 'which specifies the contra-factual conditions for a public discourse in which all limitations on time and resources have been suspended, and where the authority of the better argument prevails' (Eriksen, 2009: 45). The problem, as is usually pointed out by political scientists, and also by deliberative theorists themselves (Habermas, 1998: 323, 326), is that decision-making under the above conditions is very far from the reality of modern democracies, where the heterogeneity of individual conceptions of the good life renders rational consensuses supported by identical reasons unlikely to materialise. Hence, the problem is that if a rational consensus is not achievable under non-ideal conditions why should it be a standard with which decision-making outcomes should aspire to comply?[10] Moreover, if the rational consensus is not helpful as a normative standard, how can actual decisions then be tested? The suggestion here is to apply Eriksen's concept of a 'working agreement' instead of Habermas' demanding 'rational consensus', as the normative standard for legitimate outcomes under real-life conditions (see Eriksen, 2007a, 2007b, 2009).

A working agreement is normatively less demanding than a rational consensus, but more demanding and thus more legitimate and stable than a compromise:[11]

> an outcome may fall short of a rational consensus but still be the result of a deliberative process based on inter-subjectively justifiable reasons.... In line with this, one may think of the possibility of reaching an in-between consensus, an agreement which testifies to some movements of positions and normative learning, which does not result in a rational consensus, but in a *working agreement.* Such a conclusion rests on different, but reasonable and mutually acceptable grounds.
>
> (Eriksen, 2009: 51)

On the one hand, and in contrast to a compromise, to achieve a working agreement requires a prior process where all relevant positions and viewpoints can be presented and tested argumentatively. On the other hand, and differently from a rational consensus, a working agreement does not have to be supported by the same, but by reasonable reasons.[12] A working agreement, then, is (arguably) possible to achieve under non-ideal conditions and thus seems like a suitable *empirical approximation* of the rational consensus. But how can the political process be designed in order to increase the likelihood that decision-making outcomes have working agreement qualities?

Is deliberative democracy realistic?

Critics of deliberative democratic theory often argue that the reliance on deliberation is naïve and unrealistic, i.e. the preconditions under which deliberation can dominate the decision-making processes are so demanding that they are unfeasible under real-life circumstances (for critics, see Thompson, 2008; Bächtinger and Steiner, 2005). In short, the argument is that *power* is absent in the equation. This is a partial reading of deliberative theory, at least in the discourse theoretical variant. Discourse theory has a nuanced and multidimensional concept of power ranging from social, communicative, administrative and legal power (Habermas, 1998: Chapter 4). *Social* power refers to the capability an individual has in social relationships to assert his/her own will and subsequently to realise goals despite opposition. It reflects resources and interests in civil society, which are obviously unequally distributed. This understanding represents the traditional definition of power held by Weber and Dahl (see Habermas in Eriksen and Weigård, 2003: 172). From a democratic perspective, the problem is that social power has not been normatively assessed and will more often than not restrict the exercise of political autonomy. Unless social power is argumentatively tested both in the general publics as well as in the institutionalised settings of formal decision-making, political outcomes will only perpetuate social inequalities.

Communicative power, then, is the kind of power resulting from public deliberation, and is an expression of a cooperative and united citizenry where each citizen's equal right to communicative freedom collectively generates political power. Hence communicative power is a collective type of power resulting from a cooperative process in both the general and the strong publics (cf. Fraser, 1992; Eriksen and Fossum, 2002). The source of legitimate power thus stems from free communication. Communicative power must, subsequently, be transformed into binding and committing rules. This is done through the medium of law where political power is finally legitimately transformed into *administrative* power (Habermas, 1998: 169–70). In this regard Habermas' discourse theory relies on a two-track model of how democratic law-making can be achieved (see also Bohman, 1998: 415). The aim is to balance two spheres of power – the formal political institutions and the informal networks of society – so that they do not collapse into each other, that is, administrative power must not become self-programming and thus alienate the citizens. The constituted political system consists of a centre and a periphery. In order for citizens to influence the centre (that is, the parliament, the courts and administration) the communication of influence has to pass from the periphery through the 'sluices' of democratic and constitutional procedures. Needless to say, for the public to actually know what decision-makers are up to and thus be able to make public debate relevant for actual decision-making, *publicity* obviously represents an important precondition in order to facilitate a public debate of a certain quality and enough critical bite.

For the realisation of this system of democratic decision-making, deliberation as the mode through which decisions should be taken, is consequently inherently tied to the existence of law which is the binding element between communicatively reached decisions and their actual implementation. And, vice versa, *democratically legitimate law* cannot do without prior communicative processes where arguments have been tested and scrutinised. Moreover, as Gutmann and Thompson (2004: 4) put it: 'Assertions of power and expressions of will, though obviously a key part of democratic politics, still need to be justified by reason.' In other words, what critics often miss in their evaluation of deliberative theories is the relationship between these dimensions of power and particularly how democratic deliberation and law mutually presuppose each other in the realisation of democratic government. The reliance on law obviously makes deliberative democracy more realistic, but its communicative aspirations are still quite demanding. The challenge is to identify mechanisms that contribute towards establishing and enhancing the autonomy of deliberative settings and thus also contribute towards compensation for social inequalities and entrenched power structures at the institutionalised level of formal decision-making processes. In other words, the aim is to maximise the likelihood that social power is translated into communicative power before it becomes administrative power. How can settings for the communicative testing of policy proposals before they turn into law (as in the OLP) be realised and organised in practical terms?

The need for egalitarian law structures

Given the lack of a traditional nation-state democracy at the European level, other strands of deliberative theory part with the discourse theoretical variant defended here. In addressing the prospects for a democratic and deliberative model in the EU, defenders of directly-deliberative polyarchy hold that the Union's democratic deficit can be mended through a different model which does not possess the standard characteristics of the state, that is, it is neither sovereign nor hierarchical (see Cohen and Sabel, 1997, 2003[13]). Rather, it is capable – through the mechanism of deliberation – of ensuring intelligent problem-solving and decision-making by involving a plethora of different actors in committees, networks, publics etc. at the European level. The actors represent both private and public interests such as business experts, governmental interests, civil society organisations, NGOs etc. and none of them are capable of dominating the processes and outcomes completely. The EU is consequently seen as an advanced inter-institutional 'checks and balances' system where there is no absolute power and which ensures deliberative inclusion and testing in many forums and on many levels. This system of deliberative governance can then function as a democratic equivalent on the supranational level.

However, and unlike proponents of directly-deliberative polyarchy, discourse theory assumes that the prospect of actually transforming social power into communicative power in a democratic sense is not possible without the existence of representative institutions and formal legal arrangements for adjudication, execution and implementation in whose absence political equality cannot be assured. At best, directly-deliberative polyarchy can only ensure *deliberative governance* not *democratic government*, that is, 'authorized rule through accountable and popularly elected institutions for policy-making and implementation under the supervision of courts' (Eriksen, 2009: 177). It is not enough that a decision reflects the 'acquisition and employment of knowledge and the force of reasons' in the limited sense of providing a setting where the mode of interaction is dominated by arguing. It rather means that decisions have been subjected to the *public* use of reason and hence satisfy the basic right to justification that all affected parties by definition possess, i.e. congruence between the addressees and authors of the law.

Consequently, from a discourse theoretical perspective the quality of decisions not only depends on their epistemic value in a limited sense, but also in an extended version depicting the relevant knowledge brought to the debate by all affected parties. The only way to ensure that affected parties are included is to provide them with rights to be exercised in a democratic procedure, protected and founded in an egalitarian law-making system which has institutionalised settings where arguments can actually be discussed in this way. Only when affected parties have obtained rights enabling them to become co-authors of law can the (pragmatic, ethical–political and moral) quality of decisions be tested and trusted. The bottom line is that deliberation in itself is no guarantee for democratic government:

> deliberative governance cannot bear the burden of legitimacy as there is no possibility that all can participate on an equal basis, so that the laws that they have to obey could be consented to in a free, open and rational debate by all the affected parties. It compensates for the lack of influence brought about by globalization but is *no substitute for democracy.* When deliberation is seen as a cooperative activity for intelligent problem-solving in

> relation to an independently defined cognitive standard, as is the case with DiDeP [i.e. directly-deliberative polyarchy], it is not an argument about what is correct in the sense that it can be approved by everyone.
>
> (Eriksen, 2009: 175)

The institutional model defended by directly-deliberative polyarchy is thus not capable of distinguishing properly between when deliberation is democratic and when it is merely cognitive-scientific/technocratic. The outcomes resulting from such a system are, subsequently, also not *democratically* trustworthy even if they may have been reached deliberatively.

The solution to this problem, then, is not unique to deliberative democrats as they draw on the already well-established *practices* and *principles* underpinning the modern democratic Rechtsstaat (cf. Habermas, 2006: 412–3). It is no coincidence that this model has been so successful – its viability is based on a strong emphasis, on establishing and maintaining checks and balancing mechanisms such as a basic rights catalogue (citizens are given participatory rights also against political authorities), the separation of executive, legislative and judicial powers, a competence catalogue, the separation of state, market and civil society and the introduction of separate decision-making, accountability-, electoral- and representation procedures. These are standard mechanisms aiming at ensuring that not only the strong and powerful, but also the weaker and less resourceful groups, get to influence decision-making processes. In a deliberative reading, the purpose of these mechanisms, then, is to induce – to promote and protect – argumentative behaviour as far as possible. Hence, deliberative democracy should be seen as a supplement to liberal democracy, not as a replacement.

In bringing up the democratic Rechtsstaat as a model, some readers will perhaps question whether democracy is achievable beyond and without the apparatus of the nation-state. Whilst this is obviously a pertinent issue in this context, I will not discuss this topic here, but contend that it is (arguably) possible to transfer some of its basic principles and mechanisms/practices (such as inclusion, publicity, accountability, separation of powers etc.) to the European level.[14] The point is that regardless of constitutional framework, political institutions aspiring for democratic legitimacy should, according to the deliberative perspective, be judged 'by whether they serve to enable the medium of deliberation, which is in turn valued because of its intrinsically desirable normative properties relative to other possible means of doing politics' and subsequently whether it can be argued that the outcomes it brings about were reached in a process that comply with basic democratic procedural standards (Warren, 2007: 274). This point is equally valid for EU institutions and will be the departure for the analytical framework developed below for the assessment of the OLP. What, then, can be learned from other studies aiming to operationalise deliberative theory?

Deliberative legitimacy versus democratic legitimacy

It becomes clear from the above, that there are certain pitfalls that should be avoided when the aim is to approximate deliberative democracy to empirical conditions and that this pertains first and foremost to the flawed assumption that deliberation as such can automatically be equated with democracy. As argued above, there is a major normative difference between the legitimacy that stems from a deliberatively reached agreement among experts and technocrats versus the legitimacy that stems from a deliberatively reached agreement among popularly elected representatives defending and justifying decisions in a publicly accessible manner such as a parliamentary setting. The first can claim respect for its assumed cognitive-scientific superiority and/or efficient way in dealing with a difficult problem, but it does not deserve respect because it is democratically legitimate. Only the second type can claim respect for its putative democratic legitimacy because its validity has been tested in a publicly accessible manner. In other words, deliberation yields different types of merits, which we can label under the category *deliberative legitimacy*, that is, problem-solving, effectiveness, efficiency, cognitive-scientific quality. As we can see, deliberative legitimacy is not the same as democratic legitimacy as there is a major normative difference between outcomes resulting from either of the two types of processes.

When deliberative democracy is operationalised and applied to empirical studies this difference is, as argued above, not always taken heed of properly and thus results, in my opinion, in many misunderstandings.[15] With regard to the EU, the study of Gehring and Kerler (2008) is interesting and at the same time unclear, as they are seemingly aware of the difference between democratic legitimacy and technocratic legitimacy, but nevertheless end up confusing the reader.[16] From the outset, their study seems specifically helpful in offering a 'blueprint of an institutional arrangement that promises to stimulate deliberative decision-making, thereby producing legitimate and problem-adequate regulatory decisions'. This can be achieved if the procedure is designed in such a way that it separates 'the two tasks to be fulfilled in a deliberation, namely the definition of the broader lines of governance and their specification in concrete situations' (Gehring and Kerler, 2008: 1003, 1019). In other words, the key institutional mechanism is to ensure a division of labour between those actors setting the overall principles and goals (political level), on the one hand, and those applying these goals and principles to particular cases (technical level), on the other. The separation of the two tasks will, according to the authors, help to neutralise actors' incentives to bargain and rather gear them towards deliberation instead. The fulfilment of deliberative legitimacy thus requires both political and technical/expert legitimacy (ibid: 1018–19). The problem with Gehring and Kerler's argument is, however, that while they are steeping the article in a democratic jargon (cf. references to 'affected parties', 'democratic deficit' etc.), they are not really talking about democratic legitimacy but deliberative legitimacy, where the goal of the latter is first and foremost to ensure that decisions are reached deliberatively rather than through bargaining:

> Collectively binding decisions may be considered legitimate if they emerge from deliberation, or if they are based upon a deliberative rationale. Their legitimacy is based upon the fact that they have evolved as the most convincing solutions from discursive exchanges of reasonable arguments. To create legitimacy, a deliberation must consider all possible convincing arguments that might have an impact on the collective decision, but does not necessarily have to include all those possibly affected by the decision. Deliberative legitimacy will be undermined if the decision is reached by resources other than reasonable arguments, such as threats, promises, and package deals or other forms of horsetrading.
>
> (Gehring and Kerler, 2008: 1018)

As we can see, deliberative legitimacy is not the same as democratic legitimacy. Gehring and Kerler are not primarily concerned with establishing the conditions under which democratic decision-making can be ensured. They are simply taking for granted that the current system where the Commission, the Council and the EP take the political decisions can automatically 'mobilize "democratic legitimacy"' (ibid: 1003). This political level of the EU decision-making system is at no point in the article further problematised. They are rather concerned with arrangements where decisions can be taken deliberatively and where bargaining incentives can be kept to a minimum. Deliberative legitimacy is thus concerned with minimising the space for bargaining and as far as possible hindering decisions from being taken through other means (e.g. bargaining) than deliberation.

Democratic legitimacy, on the other hand, is first and foremost concerned with advancing institutional arrangements that can ensure a certain congruence between decision-makers and those affected by them. In this equation, deliberation obviously plays an important role, but democratic government is not co-extensive with, or limited to, deliberation alone. Moreover, democratic legitimacy, as we shall see, is not incompatible with bargaining as such, as long as it is conducted under conditions that have been democratically authorised in advance.

So even if the authors are aware of the difference between deliberative democracy and technocratic authoritarianism, their approach is not analytically equipped to take heed of this foundational normative difference as their institutional arrangement does not underline clearly the normative hierarchy between popularly elected representatives and unelected experts/bureaucrats (ibid: 1019). In short, their theoretical apparatus cannot account for the distinction between democratic legitimacy and deliberative legitimacy.

Like Gehring and Kerler, Steiner *et al.* (2004) also do not distinguish sufficiently between deliberation as a theory of human interaction and as a theory of democratic legitimacy.[17] They too collapse the difference between deliberative legitimacy and democratic legitimacy. These authors are not concerned with the EU as such, but they look at deliberation in institutionalised decision-making settings – that is, parliaments – and are therefore of immediate interest for my study.[18] Their research design is more comprehensive than Gehring and Kerler's as it contains both their (now quite well-known) 'discourse quality index' which measures the quality of actual discussions/speech acts, as well as a list of context-specific or institutional indicators that, they claim, stimulate deliberation. This is clearly one of the strengths in the approach of Steiner *et al.* For this study, it is first and foremost the latter that are of interest as the aim here is to assess the procedural qualities of the OLP not the actual discussions of concrete dossiers. They offer a list of five institutional antecedents that, they assume, spur deliberation, and hypothesise that deliberation is more likely to occur:

1 in consensus as opposed to competition democracies;
2 when there are strong- as opposed to weak veto players;
3 in presidential as opposed to parliamentary systems;
4 in second as opposed to first chambers;
5 in non-public settings as opposed to public settings.

Also, Steiner *et al.* steep their contribution within the framework of deliberative democracy, but, as mentioned, it is not entirely clear whether they merely look for institutional arrangements that spur deliberation as such or whether they aim at institutional arrangements where deliberation is conducive to democratic decision-making. The main problem is that they both do and do not treat deliberation as a normative concept distinct from its action-coordinating status. The confusion becomes discernible when they discuss consensus versus competition democracy and strong versus weak veto players (ibid., pp. 79–84). Here they only consider deliberation among elites, not whether parliamentary bodies are sufficiently open and inclusive towards input from the general publics, i.e. they take as a given that this is the case. Moreover, the hypothesis that non-public settings are more likely to spur deliberation than public settings is not followed by a discussion on if, and when, closed settings are legitimate from a democratic point of view (ibid., pp. 87–8). In other words, the criteria presented in Steiner *et al.* are not equipped to distinguish between a *democratic* deliberative decision-making procedure, as opposed to merely a deliberative decision-making procedure. The reason for this is primarily the lack of a criterion that can account and check for whether popularly elected representatives have earned/still deserve their status as representatives of citizens/affected parties when they revert into closed settings.

Moreover, they underline the existence of grand coalitions as a precondition for inducing deliberation in consensus systems because when 'they are made permanent and parties obtain relative secure power positions, the importance of electoral competition among them is reduced' and they also contribute to 'obscure policy making clarity' (see Bächtinger *et al.* 2005: 226–7). It is highly questionable whether these qualities of grand coalitions will pass a democracy test as they hinder the transparency of policy alternatives and policy disagreements. In sum, the point is that what they are measuring/evaluating is not primarily democratic legitimacy, but deliberative legitimacy. The institutional mechanisms that are more likely to induce deliberation (according to Steiner *et al.*) cannot, however be used automatically and at the same time, because indicators of democratic legitimacy as democratic government (even in the deliberative democratic version) are agnostic towards choices between consensus versus competition systems or majoritarian versus presidential systems etc. Whereas it seems more plausible to argue that consensus systems are more likely than majoritarian systems to enhance deliberation (rather than bargaining), it does not seem equally plausible to argue that consensus systems are automatically more democratic than majoritarian systems, especially when consensus systems have a tendency to hide conflict.

What these studies show, then, is that we should be careful to describe the situations and conditions under which various conclusions about the theory of deliberative democracy are reached in order to capture the normative variation that empirical findings uncover: 'While claiming (correctly) that deliberative theories share a common core of values, the empirical studies actually adopt diverse concepts of deliberation and examine different consequences under a range of conditions' (Thompson, 2008: 501). As with the EU studies, there is a need for a more nuanced approach that respects the normative difference between deliberative legitimacy and democratic legitimacy. In my opinion this is the main problem when the theory of deliberative democracy is subjected to empirical testing. It blurs the distinction between the assumptions of deliberation based on social theory (i.e. communicative action) versus the normative theory of deliberative democracy. Whereas the aim of the former is to map, describe or explain action coordination, the purpose of the latter is to establish if and when deliberation can contribute to democratic decision-making. There is also a difference between evaluating the quality of deliberation and evaluating the quality of democratic legitimacy. A main argument in this book is that in order to determine the latter it is not enough to look at what has been said and discussed – i.e. the deliberative sequences among the participants. Whether deliberation and deliberative processes contribute to democratic legitimacy depend on the context in which they occur.

Another interesting study seeking to apply deliberative democracy to empirical conditions is conducted by Nanz and Steffek (2005). Their focus is somewhat different than Steiner *et al.* in the sense that they establish criteria for the assessment of the potential democratic quality of deliberation in international governance bodies. However, as opposed to the other studies mentioned above, Nanz and Steffek distinguish clearly between elite versus democratic deliberation, and hence also between democratic legitimacy and other sources of legitimacy. The purpose is to establish empirical indicators that can 'measure' the level of inclusion and influence that the views of affected parties have on international decision-making. Since international decision-making structures are even more remote from ordinary citizens than national bodies, Nanz and Steffek hold that the arguments of affected parties are usually (and perhaps most effectively) voiced through CSOs.[19] Hence these organisations function as democratic intermediaries voicing the concerns of affected parties at the international level where democratic institutions are lacking. Their point of departure is therefore that it is possible to assess the level of inclusion of affected parties by investigating the degree to which such organisations are included and heard as participants in international governance structures. Hence they also evaluate procedural preconditions and not actual deliberations (apart from the criterion of responsiveness). The democratic quality of international deliberations is consequently evaluated according to whether civil society organisations: (1) get access to deliberations; (2) whether the policy process is characterised by transparency and access to information; (3) responsiveness to stakeholder concerns; and finally (4) whether all voices are included.

Though Nanz and Steffek's criteria resemble my own (see Chapter 3), their evaluative target is different as they define the democratic quality of international organisations according to whether CSOs are included as participants in international policy-making deliberations. This book, on the other hand, has a different starting point in the sense that it is concerned with polities that at least comply with a minimum of what is expected from democratic institutions. From their article, it is unclear whether their framework includes a political organisation such as the EU as they only mention WTO, UN and ILO as representatives of the kind of international organisations they have in mind. In contrast to these organisations, the EU has a directly elected parliament and has, to a greater degree, institutionalised procedures and processes for ensuring the inclusion of citizens' rights (e.g. Union citizenship, Charter of Fundamental Rights, ECJ, Ombudsman, treaty provisions on openness). Hence it is not the CSOs, but the elected representatives who are positioned to voice the views of affected parties. The possibility of discharging decision-makers through elections provides Union citizens, therefore, with a greater potential for holding decision-makers accountable. In addition to how the views of affected parties are included, I also evaluate the democratic quality of how competent parties are involved as well as whether the organisation of decision-making procedures more generally promotes or hinders democratic decision-making. This is important in the EU as even if CSOs may provide a link between citizens and decision-makers and thus contribute to the establishment of a European public sphere, to reduce the assessment of the participation of affected parties solely to the inclusion of CSOs is not enough in order to assess the level of inclusion in EU decision-making processes. The Union has constitutional traits and the scope of its tasks is more comprehensive and wide reaching. In terms of assessing the democratic qualities, these aspects set the EU apart from other international organisations.

Conclusion

In this chapter, I have presented the discourse theoretical version of deliberative democracy which is defended in this book. The first part of the chapter was devoted to how the deliberative perspective – in contrast to bargaining and aggregative perspectives – approaches the question of democratic legitimacy. This basically amounts to ensuring communicative testing of policy proposals in decision-making processes, where the views of all those affected stand a chance of influencing the discussion. I argued that this warrants a system that ensures political equality and thus egalitarian structures of law-making.

In the second part, the focus was on the significance of the normative difference between deliberative legitimacy and democratic legitimacy. This difference is often not taken heed of in empirical studies applying the deliberative perspective, which results in misunderstandings regarding the normative standing of the findings as well as the subsequent conclusions that are reached about what deliberation can and cannot do. The task of the next chapter, then, is to develop an evaluative scheme that takes into account these distinctions and that specifies the conditions under which decision-making processes can be assumed to warrant respect for being reached in a democratic legitimate manner.

3 Approximating ideal principles to non-ideal conditions

The evaluative framework

Introduction

In this chapter I develop an evaluative scheme, which can be employed in the assessment of the OLP. The aim is also to contribute to the debate on the feasibility of deliberative democracy by discussing and making explicit the institutional and procedural conditions that are deemed necessary for its realisation. Hence the problem formulation: *Are the institutional and procedural qualities regulating the ordinary legislative procedure organised in such a way that the process of decision-making warrants the presumption of acceptable and thus democratically legitimate decisions?*

More specifically, I develop an *evaluative framework* for empirical studies of democratic legitimacy in *institutionalised decision-making settings*. The evaluative scheme is based on a discourse-theoretical version of deliberative democracy.[1] To concretise the necessary preconditions, I have synthesised and substantially developed criteria from Eriksen and Skivenes (2000), Eriksen and Weigård (2003, Chapter 10) and Eriksen and Fossum (2002).[2] In short, a deliberative-democratic decision-making procedure must:

1 facilitate *democratic deliberative meeting places*;
2 *include* the viewpoints of *affected* and *competent* parties;
3 take decisions in *openness* so that the relevant information and documents are accessible and the opportunity for public debate and scrutiny are possible;
4 provide structures and procedures for *neutralising* and balancing asymmetrical power relations;
5 have *decision-making* capacity.[3]

The chapter has two main parts. In the first part, the normative criteria are discussed, including a section towards the end of each criterion where the empirical indicators are presented and developed. The second part is devoted to methodological considerations.

How to approach evaluation of decision-making procedures

Eriksen and Weigård (2003) present two ways of conducting a discourse- theoretical evaluation of the democratic qualities of a polity's institutional and procedural structure. The first is to assess *argumentation procedures*. Eriksen and Weigård (2003) suggest following Alexy's process-, validity- and justification rules of argumentation. See Alexy, (1989: 188, 193, 169f). Here, the first intention would be to study how interaction actually takes place by following concrete policy processes in order to find out whether they are in compliance with the discourse principle. Just to illustrate: An evaluation of argumentative procedures could for instance be based on (1) an assessment of deliberative processes from the perspective of the five types of discourses (pragmatic, ethical-political, moral, negotiation or legal); (2) an assessment could be based on an empirical classification of the type of phenomenon, and how it is understood and perceived; or (3) it could be a dissection of arguments in terms of detecting inconsistency and perversion.

The second, which is the task of this study, is to assess the *formal procedures*. This means looking at the institutional and procedural *mechanisms* and *preconditions* (e.g. meeting places, group compositions, mandate, decision rules, complaint mechanisms, openness, power neutralisation etc.) necessary for the realisation of the discourse principle. The democratic Rechtsstaat testifies to the fact that institutions and procedures matter, and that they empower as well as limit certain types of actions in a desired direction. However, formal principles do not usually provide clear-cut recipes for action, as rules in themselves never determine action. Consequently, the aim is not only to discuss the formal rules and design of the OLP as regulated in the Treaties but, in addition, to see how the formal rules and design are 'interpreted' by the actors into practices and patterns where informal norms are also important for the assessment of the democratic quality of the OLP.[4] The point is to determine how formal rules and informal norms favour or disfavour the process of democratic decision-making under the OLP. Just as formal rules open up or limit the possibility for action, so do informal norms and practices (Farrell and Héritier, 2003a, 2007).[5]

The criteria for democratic legitimacy developed below reflect a discourse-theoretical reading of the principles and procedures known from the democratic Rechtsstaat. This means that whereas the criteria are well known from other theories of democracy and in this sense are not 'new', they are interpreted and emphasised from a deliberative point of view. For instance, the first criterion of democratic deliberative meeting places underlining the importance of actual settings where policy proposals can be tested and justified in openness, will probably not figure on top of the list of aggregative theories of democracy.

What is more, in the following, I make a (strong) distinction between elected politicians and unelected experts/bureaucrats. This distinction represents a central principle of democratic constitutional states, which justifies the separation of powers in general and the normative hierarchy between democratically elected legislators and bureaucrats in particular. Thus the democratic Rechtsstaat defines and discriminates between various institutional spheres where different types of actors and tasks find their legitimate roles.[6] In other words, unelected officials have a justified place in decision-making, but when the task is to set norms and goals and to decide on the (re)distribution of goods and burdens into binding law, elected representatives ought to be in the front seat. Scientists are needed to settle and determine questions such as 'what is the case in point?', 'what are the available options and alternative solutions among which politicians can choose?', and 'what are the likely consequences of the various choices?' In short, as long as religion is no longer a valid source of authority in settling how things *should* be done, elected politicians are the only ones that can decide how normative questions should be answered and made into binding law. Unelected experts and officials thus have a secondary (even if important) position in public policy-making. This distinction is reflected explicitly in both the first and the second criteria, but it also forms the subtext of the other three criteria. Having said this, it should also be conceded that the relationship between elected and unelected actors is usually not straightforward in real-life situations. As mentioned earlier, good decisions are also firmly anchored in knowledge that popular participation cannot compensate for. Hence the role of experts and civil servants is also obviously invaluable in democracies. The relationship between popularly elected representatives and experts is revisited towards the end of the book and is discussed more thoroughly in the last chapter.

Criteria of democratic legitimacy

First criterion: meeting places conducive to democratic deliberation

For decisions to be taken deliberatively there must be arenas where such discussions can take place. Meeting places as such can have many purposes; some are merely for discussion and preparation of decision-making, whereas others are joint discussion and decision-making forums. Others again may be designed for negotiation/bargaining, aggregation/voting or conciliation. From a deliberative perspective, however, it is vital that there are some settings that are conducive to *democratic* deliberation. In other words, the relevant decision-making bodies must provide forums, organised and regulated in such a manner that they facilitate a publicly accessible debate where the arguments and positions relevant for OLP dossiers are presented, scrutinised and tested before a final decision is taken. This is important for two reasons. First, arguments and positions need a quality check, that is, they must be tested and scrutinised in order to be deemed publicly acceptable. This alludes to the epistemic dimension. Second, citizens must be assured that their concerns, interests and preferences have *in fact* been raised, considered and justified during the decision-making process (and if this is not the case, they can – on an informed basis – criticise this in public debate). Hence the moral dimension respecting the inclusion of affected parties:

> Even measures that have been agreed to by impeccable democratic procedures are likely to be seen as forms of arbitrary domination, if they are not

> also accompanied by a public forum that allows those who would have preferred alternative outcomes to see for themselves that their views have been argued and reasons given for setting them aside.
>
> (Lord, 2007a: 147–8)

In short, this is the approximation of how representative systems can ensure 'the general accessibility of a deliberative process whose structure grounds an expectation of rationally acceptable results' (Habermas, 2001: 110).

The task here is therefore to find an empirical method of singling out those settings that have the potential for ensuring such a democratic deliberative test of OLP dossiers. There are two important elements characterising this type of setting: it must be (1) deliberative, and (2) done in a democratic way. We should consequently distinguish analytically between *democratic* deliberative meeting places where the views of citizens/affected parties are included, and *expert* deliberative meeting places where only the views of competent parties/experts meet to discuss matters relevant for decision-making.

In other words, a *democratic* deliberative meeting place is not just any kind of setting. Rather, the nature of a democratic deliberative meeting place is *political* – it is an *open* and *public* setting where popularly elected representatives engage in discussions on matters of common concern[7] and where policy alternatives are formed and articulated. The citizens do not participate directly, but their 'presence' is reflected in their role as audience – a third-party judge towards whom decision-makers are accountable by the duty to present and justify their positions in a manner that is not only oriented towards their political opponents (a dyadic relationship), but which could also be deemed convincing if all those affected *were*, in fact, physically present (a triadic relationship). Moreover, the influence of the public as audience is also linked to the fact that it is periodically empowered to replace decision-makers if the former do not want the latter to continue as their representatives, that is, the politicians are constrained by the 'anticipation of retrospective control' (cf. Elster, 1998b: 2). Democratic deliberative meeting places are therefore manifestations of how and where the *views* of citizens/affected parties are included.

This means that even if a discussion forum of experts can be deemed deliberative in the sense of being dominated by an argumentative mode of interaction, it does not qualify as a democratic deliberative meeting place since it lacks the inclusion of citizens and hence the democratic aspect of deliberation. In other words, the *type of actors* is highly relevant for the definition of a democratic deliberative meeting place.[8] Moreover, through the above type of discussion positions are not only tested and scrutinised, but decision-makers are also engaging in a process of *collectively* moulding a common will. When will-formation processes in institutionalised settings are organised in this way, it (arguably) opens up for a situation where decision-makers are learning, being convinced and potentially changing their position during the process. As argued earlier, agreements reached in this way are more stable and legitimate than those which merely go through a process of aggregating individual (or party political/national) preferences. The archetypical theoretical example/ideal of a democratic deliberative meeting place is naturally the parliament and this is part of the reason why we may argue that there is a parliamentary bias in deliberative democracy.

However, as we have seen, the cognitive–scientific element of policy-making is highly important in order to reach rational decisions, and expert discourses thus have a natural place in the legislative process. Expertise here covers scientific, technical, procedural and legal knowledge, but excludes the type of knowledge affected parties have about their own situations, desires and needs, which is relevant when decision-makers make political choices about the (re-)distribution of burdens and goods and which goals and norms to pursue.[9] Consequently, an *expert* deliberative meeting place refers to forums where competent parties/expert participants *formulate, and exchange* views in an argumentative and reasoned manner. Such a setting does not have to be as public and political as a democratic deliberative meeting place, but it is not completely exempt from such requirements.

The point is that in legislative processes, expert deliberative meeting places are subordinate to democratic deliberative settings in the sense that actual decision-making is not supposed to take place in them. Opinions, arguments, criticism and concerns should be raised, aired and forwarded in order to prepare the ground for the formulation of policy proposals, but since an expert deliberative meeting place cannot be equipped with decision-making capacity it can afford to be less public and less political than a democratic deliberative meeting place. It is less political in the sense that the discussions are focused on scientific and technical information rather than assessments of moral and ethical-political considerations. The discussions are, however, political in the sense that they are related and relevant to the given policy proposal in terms of pointing at possible dangers, consequences and outcomes connected to different alternative avenues of such proposals. It is less public in the sense that it does not have to be responsive to the arguments from the wider public sphere. It has, however, a publicity responsibility in making its reports, deliberations and opinions easily accessible to all interested parties. This means that meetings may be held behind closed doors, but the minutes and information about the participants of the meetings should be available. In other words, it must be possible for the public to examine the assumptions behind, and the validity of, expert recommendations/solutions.

In *analytical* terms, deliberative meeting places (democratic and expert) are, then, distinguished from *voting/aggregative* and *bargaining* meeting places. In a purely voting/aggregative arena there is hardly any prior deliberation or bargaining and this does not, therefore, provide for the possibility of collectively moulding a Common Position. It is rather designed for the mapping and/or counting of various preferences. This is more or less also the situation in bargaining settings although here there is a longer process of exchanging positions/threats before a decision is finally taken. In a bargaining setting, the actors get acquainted with each other's positions, but it is the relative strength of the various actors that decide the outcome. In other words, in both voting/aggregative and bargaining settings a process of *collective* opinion- and will-formation is missing, opinions and arguments do not need to be tested. The direct outcome of a voting setting is thus only an arithmetic fact signifying the majority at a given point in time, whereas the direct outcome of a bargaining setting is a compromise. It is only deliberative settings that can produce more consensual outcomes.

At this point we should keep in mind that discourse theory neither expects deliberation all the way through the decision-making process, nor does it dismiss bargaining as such. Rather, it qualifies *when* and under *what kind of circumstances* bargaining is acceptable. If it follows after one or several settings where the pre-conditions for democratic deliberation are maximised, institutionalised bargaining settings where issues can be finalised within a shorter time frame, are legitimate. *Legitimate* bargaining settings are therefore formally regulated, dominated by a representative selection of elected decision-makers and they appear after one or several settings that are regulated for discussion. When bargaining is about the distribution of goods and burdens and when it occurs in a decision-making framework where the end product is binding law, the actors must be elected representatives. From a discourse-theoretical perspective, competent parties do not have the normative authority to bargain for settling outcomes with binding effect. Moreover, a legitimate bargaining setting can be closed during session in order to more easily facilitate agreement among the participants, but the discussions must be recorded and become available to the general public after the meeting is over.

Here, however, our task is to look for democratic deliberative meeting places. The indicator for such settings is that there is at least one setting per legislative house (i.e. the EP and the Council) that fulfils the following requirements:

- The setting is dominated by (a representative selection of) elected representatives.
- Access to the documents used and discussed in the setting is readily available in electronic format; discussions are conducted in openness by securing (translated) verbatim records and preferably (live and stored) internet transmission of meetings; the (verbatim) records identify who voted and what (if any) votes were cast.
- There are rules for ensuring and inducing equality among the participants: Rules should be (formally) described in the Treaties and/or rules of procedures/inter-institutional agreements stating the purpose of the setting and confirming that elected representatives are the main actors. The rules must describe the competencies, rights and duties of the setting. There must be rules for when and how the meeting ends (e.g. voting) and a commitment to announce the meeting agenda to all involved actors at the same time. There must be rules, governing debate and argument, which respect the principle of equality among the participants (unless there are good reasons given for why this should not be so).
- Time for debate and judgement prior to decision-taking is allocated.

Second criterion: inclusion of affected and competent parties

From a discourse-theoretical perspective, democratically legitimate decisions have – as discussed above – both a moral and a cognitive element, i.e. they should both be 'right' and 'true' in the sense that they are founded on the arguments of both affected and competent parties. In institutional and procedural terms this means that decision-making bodies must be able to include the viewpoints of those individuals who are going to be bound by the decisions as well as the relevant expert perspectives as mentioned under the previous criterion. Let us first look at the inclusion of affected parties.

It is the degree to which a polity is capable of including affected parties that represents the real quality test, because there is no basis for *democratic* policy-making unless citizens are not somehow recognised as rights-holders/co-authors of binding law (Habermas, 1998). Consequently, the requirement to include affected parties is at the same time both the most important and the most difficult criterion to satisfy, for how do we know that all relevant views have been heard and included in the debate when modern democracies are based on representative systems? How can citizens be recognised as co-authors of decisions when they are never present to make them?

We have already touched upon this problematic, but in the literature there are different answers to this question. With regard to the level of broad-based participation or direct citizen involvement outside general elections, deliberative theory parts significantly from republican-inspired theories defending direct/participatory democracy. Unlike the latter, discourse theory is not based on the requirement that everyone has to participate in decision-making, but rather that everyone's views, needs and interests have access to the debate as it is neither possible nor (arguably) desirable in large and complex polities that everyone should participate directly (Eriksen and Weigård, 2003: 210–2; see also Gutmann and Thompson, 2004: 31).[10] In discourse-theoretical terms, the point is not the number of people who participate, but how well all views are represented in the discussions, i.e. the 'representativity of opinions' (Eriksen and Skivenes, 2000: 28).[11] In fact, too many participants can disturb and obstruct the process and hence also the quality of the debate under non-ideal conditions (Eriksen and Weigård, 2003, 210–2). The key argument is to include the plurality of needs, interests, preferences, facts and positions so that an as qualified as possible decision can be made.[12] Hence as underlined above, discourse theory relies on the existence of satisfying procedures (Habermas, 2001: 110). That is, in a situation where it is physically and realistically impossible for everyone to participate, democratic decision-making warrants suitable and adequate arrangements, provisions and rules for the (s)election of decision-makers, for how they take decisions and for how they can be held to account.

Since discourse theory does not rely mainly on direct participation from citizens it is consequently, and as argued above, not unimportant *who* participates in the deliberative process.[13] The reason for this is due to the normatively important difference between *democratic deliberation* and *expert deliberation*. Just because actors involved in decision-making adhere to the logic of arguing this does not necessarily mean that these decisions can be deemed democratic, that is, the actors must have certain qualities. In other words, the gap between affected parties and decision-makers must be squared by 'the parliamentary principle of establishing representative bodies for deliberation and decision-making' (Habermas, 1998: 170).

Again, the parliamentary bias in the discourse theoretical version of deliberative democracy demands that a democratic procedure entails the active participation of *popularly* elected representatives. Only they can be regarded as the spokespersons of affected parties in institutionalised decision-making settings because they can be controlled and potentially dismissed in the next election. This does not mean that all actors involved in decision-making processes must be popularly elected, it only means that they must have a central position and be 'hands-on' throughout the decision-making process. The obvious reason for this is due to the fact that public decision-making is never only about finding scientifically correct answers, but also about applying such knowledge and hence making (political and ethical) choices about what should be done and how burdens and benefits should be distributed among affected parties. Public decision-making cannot therefore be left to bureaucrats and experts alone as this will amount to nothing more than technocracy or epistocracy (Eriksen, 2009). *Democratic* deliberation, as opposed to *expert* deliberation, can only be fulfilled if it contains active participation by elected representatives.

It is, however, also important that decisions exhibit a certain cognitive quality. Expert knowledge is an intrinsic and increasingly more important part of every modern governance structure as modern societies are complex, differentiated and specialised. To include actors who have competence in scientific, technical, legal, procedural and local issues is therefore a sensible and unavoidable way of meeting this challenge. Competent parties represent a manifold group ranging from government officials/advisers, bureaucrats, scientific, technical and judicial experts to representatives of NGOs/civil society organisations. We may say that the requirement of including competent parties enhances the epistemic dimension of deliberative democracy in underlining quality and rationality in decision-making.

In sum, there is a normative hierarchy between representatives of affected and competent parties where the latter are subordinate to the former in the sense that affected parties' representatives should be the main decision-making actors whereas the competent parties have a more facilitating and advisory role in providing and mediating knowledge, viewpoints and information.

The indicators of inclusion are:

- *Decision-making bodies*: the main institutional decision-making bodies are *popularly elected* and have collective *veto* power over the final outcome.[14]
- *Hierarchy of actors*: the *elected representatives* are the *deciding actors* when *discussing and deciding* how consequences and burdens are distributed. It is required that elected representatives sit together and *collectively* mould an opinion and discuss a case prior to decision-taking. The involvement of *competent parties* is limited to preparing policy-making.

Third criterion: openness and transparency

'Transparency … is regarded as an absolutely essential component of … liberal democratic systems … *Information is the currency of democracy*' (Curtin, 1997: 23, my emphasis). The principle of openness and transparency is crucial for establishing a democratic dialogue between official political bodies and the wider public and for the latter to perform any kind of enlightened and informed critique and scrutiny of on-going legislative processes. In other words, to facilitate public scrutiny and thus actual opinion-formation in the general publics, deliberation processes leading up to decision-making must be as open and transparent as possible. This criterion thus concerns how well the political system has institutionalised 'direct' communication and accountability with the general publics in the sense of providing easily accessible and unambiguous information on actual cases. This is needed in order for citizens, civil society and other social/economic actors to have a fair opportunity to form their position and participate in opinion-formation processes in the public sphere. In the wider picture this, of course, serves the important task of establishing and maintaining a European public sphere by providing the means to free opinion- and will-formation.

The principle of openness is consequently inherently linked to democracy as a precondition for the establishment and maintenance of realistic accountability mechanisms, as well as to give citizen participation outside general elections any realistic basis. To meet this end, the principle of transparency requires that policy related documents and information are easily accessible and understandable. On a more detailed level, decision-making bodies also have a responsibility and duty to present background information about the case in question, i.e. the substantial elements pertaining to the nature and characteristics of the case. For the official political institutions this means that their rules and procedures are designed and organised to inform and communicate to the general publics the relevant aspects of decision-making processes, i.e. both procedural and content-specific aspects. This includes open meetings or easily accessible verbatim records of meetings, openly announced agendas, decision-making rules, actors involved, when and if affected parties should be consulted as well as the results from open hearings, relevant laws, regulations, cost and benefit analyses, administration practices and other policy areas that will be affected, expert reports other relevant documents and knowledge about the case. Nanz and Steffek (2005: 375) make a general distinction between background and policy documents, where the former provide 'information on an issue or problem and policy documents provid[e] information on political options and proposals.' Background documents are thus less important than policy documents and the level of transparency is higher if easy access to the latter is made possible.

Openness does not only concern the extent to which documents and information are available, this criterion also prescribes that information should be easily accessible, that is, it should easy to get an overview of the main points, what is at stake etc. (Curtin, 2007b). Sometimes it can almost appear to be an intended strategy to provide enormous amounts of information in order to overwhelm or preoccupy the public, stakeholders or opposing parties with processing all the information. Decision-making bodies therefore have a duty to present the main points, dilemmas and/or inter-related issues as clearly as possible. To achieve such a goal, a public and digitally accessible register of documents is a vital tool (cf. Curtin, 2000). In addition, competing political visions and solutions should be communicated to the public so that citizens can get an overview of the relevant choices and alternatives available. Finally, in a multilingual polity, it is also a requirement that citizens are able to express themselves and that political authorities communicate with them in a language they understand and master.

The empirical indicators of openness are:[15]

- *Policy proposal*: all institutional actors involved (and the public) can get *access* to the policy proposal before it is finally decided.
- *Access to documents*: there is a digitally accessible register of documents (or other means of ensuring easy access to documents).
- *Transparency of debates*: there are either verbatim records, video streaming of debates through the internet or minutes available in all official languages.
- *Intelligibility of votes*: there are voting records available, which include information and explanation about who voted and what position they defended.

Fourth criterion: neutralisation of asymmetrical power relations

Under this criterion I distinguish between two broad categories of power neutralising mechanisms. The first covers mechanisms that are internal to the procedure and hence detectable in the procedural set-up. I call these ex ante accountability mechanisms. The second category covers external mechanisms that kick in after the decision-making process is over. These are labelled ex post accountability mechanisms.

The ex ante mechanisms cover all those (formal, but also informal) rules that help to induce equality among participants and to induce argumentative behaviour in a manner which is as open, equal and inclusive as possible. The rules governing the OLP should be easily accessible and understandable both to the involved actors as well as the public at large. The crux here is that information about how the decision-making procedure functions – its rules, stages, applicability etc. – must be regulated and easily accessible in writing, preferably in the Treaties (describing the normal decision-making method(s)) or in ordinary statutes (special procedures). In addition, it must be predictable, understandable and consistent – that is, from the general guidelines it must be possible for citizens to get an unambiguous overview of the rules and different stages/phases of a decision-making process, but the outcomes of policy processes must not be predictable from these guidelines.

It is also mandatory that the rules have a particular content.[16] First, the procedural set-up, as well as its functioning, should reflect an explicit duty to explain and give reasons for positions and decisions towards fellow institutional members, to other institutional actors and to the public at large. This involves precise rules and regulations on when and how often, as well as what kind of setting decision-makers must attend. Second, the division of labour between the institutional actors should respect the principle of separation of government powers both informally as well as formally in the Treaties. Third, the involvement of representative institutions must be underlined in the legislative process, i.e. there must be recognition of the parliamentary principle. Fourth, the procedure should be organised so that the process has roughly the following sequencing: argumentation, potential bargaining and final decision-making. Fifth, inter-institutional agreements, as well as the rules of procedure organising the internal life of each institutional actor, should be in accordance with the above principles.

In addition to ex ante mechanisms, the democratic Rechtsstaat also offers well-functioning accountability mechanisms that can be activated after a procedure is closed. In knowing that ex post accountability measures can be triggered if an actor does not adhere to the prescribed rules, this can make them stick to responsible and argumentative behaviour rather than seizing the opportunity to deviate from such behaviour in cases where their immediate self-interest can be satisfied without adhering to arguing and justification. Ex post accountability mechanisms such as the judicial review and Ombudsman are thus mechanisms that have proved to be helpful in neutralising asymmetrical power relations.

Critics would perhaps argue – and rightfully so – that neutralisation mechanisms such as separation of powers and/or judicial review/Ombudsman can also be undemocratic. For instance, the debate on the juridification of politics and judge-made law refers to the fear of Courts being too autonomous, and acting beyond their mandate by altering – expanding or limiting – laws on their own, without any authorisation of democratically elected legislatures. While these are important debates, I will not discuss them here, but simply argue that all neutralisation mechanisms must be democratically vindicated and authorised (see Chapter 8; cf. Gutmann and Thompson, 2004: 3).

The empirical indicators are the following:

- *Ex ante mechanism* (1): *Intelligibility of the procedure:* there is a *precise description of the rules* governing the OLP. The *Treaties* contain information about the main actors and what powers, rights and duties they have. In addition, there is information about the decision-making rule (unanimity/majority vote) and the sequencing of policy stages. The institutions' *rules of procedure and/or inter-institutional agreements* are in line with Treaty specifications.
- *Ex ante mechanism* (2): *Separation of powers:* the institutional actor presenting a policy proposal is not at the same time the decision-maker.[17] There is a clear *legal division* between executive, legislative and judicial powers.
- *Ex post mechanism* (1): *Judicial review*: affected Union citizens and other legal subjects can appeal and get their complaint reviewed, on which basis

the Court can scrutinise (and possibly sanction) not only adherence to procedural rules, but also the content of OLP dossiers.

- *Ex post mechanism* (2): *Ombudsman:* affected Union citizens or other legal subjects can appeal and get their complaint reviewed, on which basis the Ombudsman has a mandate to scrutinise (and possibly 'sanction') adherence to procedural rules and can also consider the content of OLP outcomes.

Fifth criterion: decision-making capacity

To make a difference, a legislative procedure must have the capacity to produce outcomes (whatever the quality of the decision). Decision-making capacity here refers to whether the institutional frameworks formally inscribed in, and thus authorised through, the Treaties are also the institutional frameworks within which decisions are actually taken. From a deliberative democratic perspective, the answer to this criterion can give us an indication of whether the potential for publicly accessible deliberative processes that we find under the first criterion can in any shape or form be connected to, and thus be said to influence, the decision-taking moment. If we find that the OLP 'scores' well on the four first criteria, it is important that actual decision-making takes place within the framework of that procedure. Otherwise it simply rubber stamps decisions taken in other forums or by other actors. This does not mean that informal meetings are automatically deemed 'illegitimate'. Formal institutions prescribed and described in legal provisions are usually not ready-made recipes with no need for practical interpretation when applied to actual empirical circumstances. The point is that the actual practice of a procedure can be more or less in line with the *intention* of the formal legal provisions and that informal structures/meetings have not taken over and thus made the formal set-up redundant.

The empirical indicator of decision-making capacity is:

- *The absence of informal networks* where powerful actors 'pre-cook' proposals that are not subsequently sanctioned by democratic organs and which consequently have taken over as the locus where actual decision-making takes place.

Methodological considerations

Normative status and relationship between the criteria

Whereas all five criteria are important in determining *empirically* the democratic quality of decision-making procedures, it could be argued that the inclusion of affected parties is the *normative* source feeding or 'explaining' the importance of the other criteria. The reason for this is that, from a discourse theoretical point of view, the only entities possessing moral significance in and of themselves are individuals (not communities or polities) (see Stie, 2002). Hence, democracy is meant as a framework aiming at the realisation of a type of collective

Table 3.1 Criteria and indicators

1. Democratic deliberative meeting place	*2. Inclusion of affected and competent parties*	*3. Openness*	*4. Neutralisation of asymmetrical power relations*	*5. Decision-making capacity*
At least one democratic deliberative meeting place per legislative body	Elected representatives are main decision-makers and have veto power	Access to policy proposals prior to decision-taking	Ex ante mechanism (1): Intelligibility of procedure	Absence of competing decision-making structures/ groups
	Hierarchy of actors	Access to documents Transparency of debates Intelligibility of votes	Ex ante mechanism (2): Separation of powers Ex post mechanism (1): Judicial review Ex post mechanism (2): Ombudsman	

self-government that respects the moral status of all involved and affected individuals (see also Eriksen, 2009: 46–7). In this sense, openness, neutralisation mechanisms, democratic deliberative meeting places and decision-making capacity are derived from the moral status of affected parties and stand in the service of its realisation. I have also added the inclusion of competent parties as a necessary criterion of democratic legitimacy. Competent parties bring quality and knowledge into the decision-making process, but if affected parties are not included a polity governed solely by competent parties is nothing but a technocracy characterised by elite deliberation. Hence normatively, the inclusion of competent parties is also derivative of the status of affected parties.

Status and relationship between the criteria – notes on the research design

Whereas all five criteria represent distinct categories or components of democratic legitimacy on a theoretical/analytical level, we cannot expect that they will occur in a clear-cut and distinct manner when we turn to the empirical realm. The criteria are *empirically* not mutually exclusive, but rather support, complement and build on each other. It is, for instance, difficult to conceive of a decision-making procedure where there is full inclusion of affected parties, but no mechanism to neutralise power asymmetries. Or, where there is full openness, but absolutely no inclusion. For instance, to get a more reliable picture of the degree of inclusion, it is required that we also have in mind and take into consideration how well the other criteria score, as such information can qualify, modify and nuance the findings under each criterion.

What is more, criteria two to five are aimed at the overall procedure. The first criterion, however, refers to a specific but crucial component of deliberative democracy, namely the importance of ensuring that there are meeting places regulated in such a way that democratic deliberation is likely to happen. I shall argue that the emphasis on safeguarding such meeting places represents the intake to a deliberative reading of the principles and procedures on which modern democratic polities are founded as it epitomises the 'institutional structuring of incentives to deliberate' (Warren, 2007: 273). Now, from the presentation made above, the characteristics of meeting places seem more or less to be the same as criteria two to four. It is correct that to determine the democratic quality of a meeting place involves looking at its membership, its level of openness as well as whether there are procedural rules and practices that help to neutralise asymmetrical power relations. However, the properties of a democratic deliberative meeting place are not co-extensive with criteria two to four. And whereas the findings resulting from the evaluation of *individual* meeting places *can* tell us something about the *overall procedure*, it should be remembered that the findings from a single meeting place are *not necessarily* and *automatically* valid for the whole procedure. This needs to be qualified in relation to how well the other criteria apply. In other words, the conclusion reached about the potential democratic qualities of the OLP must be based on a qualified judgement of how it approximates to compliance with *all five* criteria.

With regard to democratic deliberative meeting places some further methodological comments should also be made. First, in actual decision-making processes there are usually many types of settings – some are better described as deliberative (democratic or expert) and some more accurately as bargaining or voting sites. This does not mean that the procedure under evaluation cannot be deemed democratic if not all relevant and important meeting places are deliberative according to the democratic definition. That would obviously be an unrealistic requirement. It would, however, be impossible to argue that a decision-making procedure could be democratic if none of the relevant settings had deliberative-democratic qualities. In sum, we could realistically expect instead, that democratic deliberative qualities will be on display at different junctures of the process, i.e. prior to bargaining and decision-taking/voting.

Second, it should be underlined that data on the procedural and institutional aspects only tell us so much about the deliberative democratic qualities of meeting places. We know that a setting does not qualify for the democratic category unless it contains elected representatives, but readers will most likely wonder how we can determine the *deliberative* aspect of these meeting places. This is not an easy task and it should be iterated that this type of assessment cannot give conclusive answers to the question of whether a setting is *actually* dominated by deliberation, bargaining and/or any other type of interaction mode. It can, however, point to the likelihood that deliberation will occur and one tool in this regard is to look closer at the rules for the proceedings of the meeting. Is there for instance a chair person and/or formal rules to enforce equality among the participants? (e.g. that every participant gets the chance to speak or that there is a fair balance between opposing parties to get their views heard etc.). The analytical framework can give answers to the question of whether the institutional and procedural conditions are more or less likely to enhance (or hinder) democratic deliberation. In other words, it evaluates the procedural conditions. If the setting is formally described or if it is somehow possible to find out what its purpose is, this can presumably give an indication of whether it is devoted to discussion, bargaining, voting or combinations of these. Although this is not necessarily a very reliable source of what is *actually* going on, it nevertheless provides us with information on what is *supposed* to be going on and henceforth helps us to single out those settings whose purpose is not discussion, but bargaining.

Third, whereas the above list of requirements may seem fairly clear, empirical situations are always more complex and difficult to get an overview of than theoretical models. It can therefore be harder to decide on how to categorise empirical findings, as there may well be settings that come close to, but that do not entirely comply with, all the indicators. Hence conclusions can be difficult to reach. As a rule of thumb the *setting* cannot under any circumstances pass as democratic unless the main actors are elected representatives. Being elected is, however, not enough as the representatives can lose their normatively superior status as trustees who voice the views and speak on behalf of affected parties if the meeting place is completely closed to the general public and if there is no possibility for others (either other institutional actors or the general public) to hold them accountable for their actions in this particular setting.

Relevance of the findings

As mentioned earlier, the criteria and indicators presented above do not automatically ensure or lead to democratic deliberation. It is important to underline that they do not guarantee democratic decision-making as such, but rather that they represent institutional and procedural conditions that are assumed to be necessary in order for democratic decision-making processes to take place at all. In this sense they are necessary, but not sufficient preconditions for ensuring democratic deliberation. Or put differently, they are probabilistic not deterministic. The findings and conclusions in Chapters 5–9 are consequently not able to say something about how decision-making under the OLP is carried out in each individual case, they rather tell us something about whether the institutional and procedural set-up hinders/limits or broadens/enables the likelihood of deliberative-democratic decision-making occurring under the OLP configuration. The data that will tell us about the likelihood of this taking place are drawn from other studies of how the OLP works in practice.

Critics would perhaps argue that institutional and procedural design is not the right target of analysis when the aim is to assess the OLP for its putative democratic qualities. They would maybe consider that it is the quality and content of legislative *outcomes* that really matters when we want to study democracy under the OLP. It is, of course, true that much can be learned by investigating policy outcomes. However, we should be careful about judging democratic legitimacy from the perspective of outcomes alone as even dictatorships can produce desirable decisions. In contrast, the claim guiding this book is that the institutional and procedural designs surrounding and underpinning legislative processes make a difference. Institutional designs are not put in place in an entirely random manner, since not all types of design are capable of producing democratic decisions. What is more, democratic decisions are not only about outcomes, but also about *how* they came about, hence the importance and careful choice attached to the type of procedure and institutions involved when decisions are taken concerning and affecting citizens in their daily lives. This is why modern democracies have rules for securing representation, contestation and opposition so that different political alternatives can be articulated and identifiable for the citizens.[18] Democratically legitimate decisions demanding general obedience are therefore just as recognisable for the way they have been reached, as has been discussed in Chapter 2.

However, good intentions are not enough, as institutional designs are never realised in exactly the way they are intended. To study the informal practices and processes accompanying the formal ones is therefore important in order to take stock of the (unintended) consequences institutional and procedural mechanisms open up for once they are put into practice. This is also the case with the OLP where the informal trialogue system has given momentum to the intensified informal relationship between the EP, the Council and the Commission. In particular, the EP and the Council have developed close ties and a 'shared legislative culture' with strong emphasis on informal mechanisms due to the requirement under the OLP that both houses must agree in order for a law to be adopted. To understand and to evaluate the way decisions come about under the OLP these informal settings cannot be ignored (see also Farrell and Héritier, 2003a, 2003b, 2007).

Putative normative standing of findings

The claim made here is that the extent to which the procedural criteria are complied with can tell us something about the quality of the outcomes. In this regard, we should recall the discussion in the previous chapter where the concept of a working agreement was introduced. Under non-ideal conditions where citizens themselves cannot participate in actual decision-making and where deliberation cannot go on until a consensus has been reached (or an agreement made that a rational consensus cannot be reached), some compromises must be made with regard to how rational and acceptable decisions can be expected to be. Hence citizens – as decision-makers – are replaced by elected representatives, and the deliberative expectations are lowered in the sense that the deliberative process is punctuated by temporary stops (voting) so that decisions can be taken even if a rational consensus has not yet been reached. If all the five criteria are met, then there is a basis for arguing that OLP acts have the normative qualities expected from working agreements as discussed in Chapter 2. Should that be the case, OLP agreements yield more stable consensuses than mere compromises. In other words, for empirical studies the principles that are characteristic of the ideal speech situation have been approximated to the five criteria and the types of outcomes the ideal speech situation endeavours to achieve – the rational consensus – have been approximated to the working agreement.

On evaluation

It is important to keep in mind the nature of this study. It is an evaluation according to normative standards *not* a causal analysis trying to explain or understand a certain empirical puzzle or phenomenon. By deducting a normative standard of democratic legitimacy from discourse theory, the institutional and procedural characteristics of the OLP are rather assessed for their compatibility or compliance with the theoretical criteria.

The theoretical criteria presented below are 'regulative ideals', that is, they are not expected to be fully met in an empirical analysis. That does not mean that they are inappropriate or wrong. Rather, this is exactly the nature of normative or regulative ideals/criteria as they 'are goals toward which to strive … , not standards that can be fully met.' (Mansbridge, 2003: 515).[19] Regulative ideals lose their force as normative compasses if they are not to be approximated when applied to empirical conditions. To determine the putative level of congruence between normative criteria and empirical reality therefore requires contextualisation and a qualitative discussion on how deeply the expected discrepancy between ideal and reality is to be judged. It could therefore be argued that we need three categories: ideals, standards of how close the procedural qualities of a decision-making procedure can reasonably be expected to get to the ideals, and procedural performance. The methodological consequence of this argumentation is thus in line with Mansbridge's understanding of democratic legitimacy as a spectrum rather than a dichotomy: 'Conceiving of democratic legitimacy as a spectrum and not a dichotomy, one might say that the closer a system or representation comes to meeting the normative criteria of democratic aggregation and deliberation, the more that system is normatively legitimate' (Mansbridge, 2003: 515). Hence, aided by the operationalisation of the abstract democratic ideals, the evaluative framework providing a specification of concrete criteria and indicators facilitates the means to determine the threshold of procedural compliance a decision-making procedure should reasonably be expected to meet in order to be deemed democratic. Or to put it in the European context, then:

> It goes without saying that democracy is always unfinished. It is a contested concept and an ideal that can never be fully realized. We can therefore only talk about *democratization* of established power structures as a measure of democracy. Thus the end product is not democracy *tout court* but a state of affairs that complies better with proper standards than the present state of affairs in Europe. By democratization we generally mean curtailing the level of domination in society and increasing the possibilities for collective self-determination. In line with the theory of deliberative democracy, democratization is here further specified to mean increasing the possibilities for offering the citizens justifications for the power structures they are subjected to.
>
> (Eriksen, 2009: 2)

Scope and time frame

To assess a decision-making procedure in general, and the OLP in particular, is a formidable task as it can be related to almost all aspects of the EU. There is hardly any EU topic that is not in some shape or form interesting or relevant in relation to the OLP ranging from expert committee influence, lobbying, Commission agenda-setting power, inter-institutional relations between the EP and the Council (and also Commission), transparency, relationship between elected and appointed decision-makers and so forth. Consequently, to delineate exactly what should be included and excluded in an assessment of the democratic qualities of the procedure is very difficult.

Formally, the OLP is a legislative procedure confined to policy formulation/development and decision-making. The procedure is officially over when agreement is reached or, alternatively, fails. This means that, in this book, the focus is on the legislative process from when the Commission tables its proposal and up to the Council's and EP's adoption or rejection of an OLP act. In other words, the evaluation provides a snapshot into a small part of what constitutes a whole. Ideally, the whole chain of the decision-making process should be included starting from the very beginning in the Commission via the delegation and implementation procedures (cf. former comitology procedures) to the actual implementation in the member states. In addition, the structural preconditions for democratic government such as political parties, a vibrant public sphere, a common political identification as well as the connection(s) between the national and supranational levels should be evaluated in order to get a more comprehensive picture of the democratic state of affairs in the Union. However, apart from time and space limitations, confining the assessment to the legislative process proper can be justified on the grounds that it constitutes the 'most likely' case, that is, if the adoption of binding legal acts through popularly enacted bodies such as the indirectly elected Council and the directly elected EP fail to meet basic democratic requirements, what does this then tell us about the democratic situation of the Union as a whole?

This choice does not, however, make bureaucracy absent, as the process of decision-making is very much a joint effort between bureaucrats, experts and politicians (elected as well as appointed ones). One of the main aims here is to evaluate the relative importance and influence of competent versus affected parties on decision-making. Needless to say, the role of bureaucrats and experts is by no means marginal. When the objective is to assess the democratic qualities of the OLP, the task is to appraise when the involvement of unelected experts and bureaucrats is justifiable and when it is not.

With the Lisbon Treaty, the national parliaments have acquired an official role in the OLP.[20] The Commission is now obliged to send draft legislative acts to the national parliaments at the same time as they are sent to the EP, the Council, the Committee of the Regions and the European Social Committee. Through the so-called early warning system, national parliaments can formally object to OLP proposals if they find them in breach of the subsidiarity principle by means of the yellow or orange card procedures (more in Chapter 4). The early warning system does not provide the national parliaments with a right to veto OLP proposals as the EP and the Council can. Rather, it bestows upon them a formal right to have their opinion heard on the compliance of draft proposals with the subsidiarity principle, whereby the Commission is subsequently obliged to review the proposal and (if the latter disagrees with the national parliaments) it must provide reasons for why the proposal will be upheld. In other words, we could also have added the national parliaments as a potential additional ex ante accountability mechanism in their role as watchdogs of the principle of subsidiarity. Even if the early warning system could make the national parliaments more involved in co-decision-making, they are nevertheless not assessed separately as a potential democratic neutralisation mechanism due to the simple fact that only one yellow card procedure has so far been initiated by the national parliaments.[21]

The roles of national parliaments are, however, briefly presented and discussed in Chapters 4 and 10 respectively.

Moreover, the evaluation of the OLP is confined to the current legal situation, that is, the provisions of the Lisbon Treaty. This means that I will not conduct a comparison in time between the OLP now and its historical predecessors. The coming into being of the present framework is often discussed in relation to how the situation was before, but this is not the main objective and is not done systematically. A brief historical presentation of the procedure can be found in the next chapter. The OLP was introduced with Maastricht and amended by Amsterdam. The Nice Treaty extended the number of policy areas, but did not alter the basic provisions and workings of the procedure. The same can more or less be said for the Lisbon Treaty even if, as mentioned above, the national parliaments now have a formal role in the OLP. In sum, it is the Amsterdam Treaty that represents the most important revision of the procedure. I come back to this in the next chapter.

Sources and data

My account relies on an interpretation of official documents and academic studies on the OLP in particular and the EU legislative system in general. More specifically, it relies on official sources such as the Treaties, the institutions' respective rules of procedure and the joint declarations between the EP, the Council and the Commission as well as material presented on the EU's websites such as activity reports, the institutions' OLP guides, OLP online video transmissions of EP and Council sessions etc. To find one's way around the jungle of documents in the EU can be difficult and overwhelming. Apart from spending much time browsing around the EU's website in order to get acquainted with where to find information, I have therefore often used academic sources as a guide to single out which documents are (more) relevant and important.

In addition to these official sources, I use articles written by actors in the system[22] as well as academic studies by researchers of which many also contain interviews with central and relevant OLP actors.[23] I also use more general or overview accounts of the EU system to look at how the authors of these books write about the workings and mechanisms characterising the OLP.[24] What is more, I have looked for studies discussing the EU's democratic deficit in relation to decision-making processes in particular (Settembri, 2005; Farrell and Héritier, 2003b), and studies evaluating the democratic deficit in more general terms.[25]

With regard to the appraisal of the accessibility of information and documents available on the EU's website under the third criterion, the conclusions drawn are based on my own data. No studies have actually conducted detailed investigations of how much information is available for citizens if they want to follow the progress of an OLP dossier. Hence the assessment of how the three institutions' websites function as citizens' primary sources of information is conducted by the author if not otherwise noted.

Conclusion

This chapter has been devoted to the development of the evaluative scheme that will be applied to the OLP case in the following Chapters (5–9). From deliberative democratic theory, five criteria have been developed and substantiated: (1) democratic deliberative meeting places, (2) inclusion of affected and competent parties, (3) openness, (4) neutralisation of asymmetrical power relations and (5) decision-making capacity. Each criterion is accompanied by empirical indicators. The chapter ends with some methodological considerations. Here it has been underlined that the evaluative scheme is focused on the *conditions* under which democratic decision-making is more/most likely to occur and consequently not whether decision-making under the OLP is actually dominated by democratic deliberation or not. Moreover, although the analytical framework cannot determine whether deliberation (and/or bargaining) is actually taking place, it can clarify what kind of deliberation the procedural and institutional circumstances are capable of and thus likely to generate. Before delving into the evaluation proper, the next chapter introduces the OLP in more detail.

4 Background

From co-decision to the ordinary legislative procedure

Introduction

Since the coming into force of the Lisbon Treaty, co-decision has – as we have seen – become the 'Ordinary Legislative Procedure' or the default procedure so to say, and it is now the most important legislative procedure in the EU. The other legislative procedures now figure under 'special legislative procedures'.[1] The OLP is regulated in Article 294 (TFEU) and is currently applicable in most policy areas, including fisheries and agriculture, in addition to most issues formerly figuring under the justice and home affairs or third pillar.

This chapter has three parts. The next section contains an introduction to the OLP and how it generally works. Section two provides a presentation of the historical development of the OLP and section three provides some general numbers on co-decision-making.

How the ordinary legislative procedure works

Article 294 TFEU

The OLP is formally described in Article 294 of the Treaty on the functioning of the European Union:

Article 294[2]

1 Where reference is made in the Treaties to the ordinary legislative procedure for the adoption of an act, the following procedure shall apply.

2 The Commission shall submit a proposal to the European Parliament and the Council.

First reading

3 The European Parliament shall adopt its position at first reading and communicate it to the Council.

4 If the Council approves the European Parliament's position, the act concerned shall be adopted in the wording that corresponds to the position of the European Parliament.

5 If the Council does not approve the European Parliament's position, it shall adopt its position at first reading and communicate it to the European Parliament.

6 The Council shall inform the European Parliament fully of the reasons that led it to adopt its position at first reading. The Commission shall inform the European Parliament fully of its position.

Second reading

7 If, within three months of such communication, the European Parliament:
a approves the Council's position at first reading or has not taken a decision, the act concerned shall be deemed to have been adopted in the wording which corresponds to the position of the Council;
b rejects, by a majority of its component members, the Council's position at first reading, the proposed act shall be deemed not to have been adopted;
c proposes, by a majority of its component members, amendments to the Council's position at first reading, the text thus amended shall be forwarded to the Council and to the Commission, which shall deliver an opinion on those amendments.

8 If, within three months of receiving the European Parliament's amendments, the Council, acting by a qualified majority:
a approves all those amendments, the act in question shall be deemed to have been adopted;
b does not approve all the amendments, the President of the Council, in agreement with the President of the European Parliament, shall within six weeks convene a meeting of the Conciliation Committee.

9 The Council shall act unanimously on the amendments on which the Commission has delivered a negative opinion.

Conciliation

10 The Conciliation Committee, which shall be composed of the members of the Council or their representatives and an equal number of members representing the European Parliament, shall have the task of reaching agreement on a joint text, by a qualified majority of the members of the Council or their representatives and by a majority of the members representing the European Parliament within six weeks of its being convened, on the basis of the positions of the European Parliament and the Council at second reading.

11 The Commission shall take part in the Conciliation Committee's proceedings and shall take all necessary initiatives with a view to reconciling the positions of the European Parliament and the Council.

12 If, within six weeks of its being convened, the Conciliation Committee does not approve the joint text, the proposed act shall be deemed not to have been adopted.

Third reading

13 If, within that period, the Conciliation Committee approves a joint text, the

European Parliament, acting by a majority of the votes cast, and the Council, acting by a qualified majority, shall each have a period of six weeks from that approval in which to adopt the act in question in accordance with the joint text. If they fail to do so, the proposed act shall be deemed not to have been adopted.

14 The periods of three months and six weeks referred to in this Article shall be extended by a maximum of one month and two weeks respectively at the initiative of the European Parliament or the Council.

Special provisions

15 Where, in the cases provided for in the Treaties, a legislative act is submitted to the ordinary legislative procedure on the initiative of a group of Member States, on a recommendation by the European Central Bank, or at the request of the Court of Justice, paragraph 2, the second sentence of paragraph 6, and paragraph 9 shall not apply.

In such cases, the European Parliament and the Council shall communicate the proposed act to the Commission with their positions at first and second readings. The European Parliament or the Council may request the opinion of the Commission throughout the procedure, which the Commission may also deliver on its own initiative. It may also, if it deems it necessary, take part in the Conciliation Committee in accordance with paragraph 11.

The OLP step by step

The OLP is divided into three stages or readings, which can be concluded at each stage depending on the ability to reach agreement. It is common to say that the procedure begins with the Commission launching its proposal to the EP and the Council. However, prior to formulating the proposal, the Commission has conducted an extensive consultation process both inter-institutionally with the EP and the Council, externally with the civil society organisations and corporate interests etc. and also with other EU institutions, such as the national parliaments as well as the Economic and Social Committee and the Committee of the Regions, in those areas where their opinions are required. As mentioned in the last chapter, the national parliaments now have a right to forward a reasoned opinion if they find that the Commission's proposal are not in line with the subsidiarity principle and they are able to form a simple majority within 8 weeks. In other words, prior to launching a policy proposal and formally opening an OLP, the process has usually lasted quite some time in order for the Commission to have an idea of and anticipate the various positions and identify contested issues.[3] In the Commission the proposal is then finally and officially adopted either through the written (no discussion/'A' point) or oral (with discussion/'B' point) procedure among the Commissioners and then sent to the EP and the Council as well being published in the Official Journal in its 'C' series.[4]

First reading

The first reading starts when the Commission – which has the formal monopoly of initiative – launches its policy proposal simultaneously to the EP and the Council. The difference between the first reading, on the one hand, and the second and third readings, on the other, is that there are no time limits in the former. Discussions can therefore go on as long as is deemed to be necessary both in the EP and the Council. The proposal is officially announced in plenary in the EP and allocated to the relevant parliamentary committee (potentially, there is joint involvement of two or more committees if the nature of the case so requires). The task of the committee is to prepare the Parliament's 'Position'[5] on the Commission's proposal and the committee thus appoints a rapporteur who is put in charge of producing a report suggesting amendments and changes (if any) to the Commission proposal. The report is firstly subject to internal committee discussion and, when agreed among committee members (a simple majority is required), recommended to and discussed in the Parliament in a plenary session. The report is also adopted by simple majority[6] in the plenary and consequently made the 'Position' of the EP and then submitted to the Commission (as well as the Council). The Commission reviews the amendments from the EP and if it agrees, adopts a modified proposal in order to accommodate the suggestions of the Parliament.

Meanwhile, the responsible Council working party (or parties) has started the first reading process in the Council. The Council working parties are subcommittees of the Council of Ministers, which discuss, make recommendations and table amendments to the (modified) Commission proposal. The process is supervised by the members of the Committee of Permanent Representatives (Coreper) and attended by national representatives sent from the member states. The process then moves on to Coreper, which examines the agreements reached in the working parties and attempts to resolve outstanding points of disagreement before the proposal is sent to the Council of Ministers. On the basis of the EP's Position and the Commission's potential recommendations, the Council discusses the proposal and reaches a Common Position by a qualified majority vote[7] if it agrees with the recommendations made by the Commission. If the Council disagrees with the alterations the Commission has made to the EP's amendments, this requires unanimity in the Council.

If it turns out that the Council approves all the amendments made by the EP (or if the EP did not make any amendments to the Commission proposal) and the Commission, the proposal can be adopted and become law for implementation right away without the Council having to agree on a Common Position. This kind of 'fast track legislation' or 'early agreement', where the procedure can be closed after first reading, was made possible after the changes in the Amsterdam Treaty. If, on the other hand, the Council does not approve and a Common Position is adopted, this signals the end of the first and the start of the second reading.

Second reading

One difference between the first and the second reading is the shift from simple to absolute majority voting in the EP. There are no changes to the Council's QMV (qualified majority vote) rule.[8] Another difference is that during the first reading of the OLP there are no time constraints either on the EP or on the Council when formulating their Position and Common Position respectively. In the second reading, however, a strict timetable is prescribed (see below). Finally, the end of a second reading is different from a first reading sequence in the sense that the former opens up the possibility of convening the Conciliation Committee.

Apart from the above, the second reading is more or less a copy of the first. The second reading process starts with the EP taking a stance on the Council's Common Position within three months (with the possible extension to four months) after receiving it. If it approves or does not respond, the act is deemed adopted.[9] If the Parliament rejects the proposal by an absolute majority, the act fails. If, on the other hand, the Parliament makes amendments to the Common Position (supported by an absolute majority) [10] the dossier is sent back to the Council (and the Commission for its opinions). Within three months (with the possible extension to four months), the Council can either approve the EP's suggestions (by QMV if the Commission has approved them or otherwise through unanimity) or it can reject either all the amendments or just one single amendment. If approved, the Common Position with EP amendments is adopted. If rejected, the Conciliation Committee is convened within six to eight weeks. For conciliation processes, both the EP and the Council select a delegation to represent them. In the Conciliation Committee the members attempt to find a compromise position. The Conciliation Committee consists of an equal number of members from the Council and the EP with the participation of the Commission. If no agreement is reached within six to eight weeks the act is rejected. If, however, agreement[11] has been reached within the same time limit and a 'Joint Text' has been adopted, the third reading starts.

Third reading

The members of the Conciliation Committee then take the agreement back to their respective institutions where the conciliation compromise – the Joint Text – is presented and the proposal with amendments is voted upon. For the act to be adopted both institutions must approve – the EP with a simple majority of the votes cast and the Council with a qualified majority. Agreement in both houses must be reached within six to eight weeks. If one of the institutions rejects the proposal, the act is not adopted. At this stage it is not possible to open the process again and the EP and the Council must take a vote on a take-it-or-leave-it basis. A rejected act can only be re-opened if the Commission tables a new proposal.

The trialogues

In parallel with the formal procedure as described above, trialogues are now an important characteristic of the OLP. These meetings are crucial for the understanding of how the OLP works and therefore deserve some space here. The trialogue is an inter-institutional arena containing a limited number of participants from each institution. From the EP, the rapporteur, the chair of the responsible committee and, in addition, often the shadow rapporteurs (especially those from the large political groups) meet. In trialogues prior to and during conciliation, one of the three permanent EP Vice Presidents also participates. The Council is usually represented by the Presidency in the form of the chair of Coreper I (the deputy permanent representative),[12] the chair of the relevant working group and the Council Secretariat 'backbone' which provides legal expertise (Häge and Kaeding, 2007: 344; Rasmussen, 2005: 1018). The incoming Presidency is usually also included, but only with observer status (Shackleton and Raunio, 2003: 186, footnote 8). The Commission is represented by the relevant Commissioner in the trialogues that are conducted 'just before or on the fringe of the Conciliation Committee.' For the other meetings, a high-level official, usually the relevant Director-General or Director meets, assisted by the Secretariat General (the Codecision Unit and the Legal Service) (Häge and Kaeding, 2007: 344). In sum, there are not more than ten persons from each institution (EP, 2007b: 14).

In the beginning, trialogues were only common prior to the Conciliation Committee, but are today conducted throughout the procedure. Hence, whereas the Treaties formally describe the OLP in a sequential manner, i.e. that each institution deals with the dossier in turn, responding to each other's suggestions as outlined above, Farrell and Héritier (2004) point to the *simultaneity* of the procedure due to these informal practices and contacts as they take place throughout the whole procedure. In other words, it is not the case that one house sits and waits until the other has come up with a response. Rather, while the formal process proceeds in a sequential manner, selected representatives from all three institutions are in constant contact with one another in order to reach agreements, broker deals and sound out each other's positions. In 1999 with the inception of the Amsterdam Treaty and the possibility for closing the procedure after the first reading, the EP, Commission and the Council signed a Joint Declaration committing them to cooperate on a constructive basis.[13] The agreement was revised in 2007 with a particular emphasis on how the first reading trialogues should be conducted.[14] There is now a strong emphasis on reaching so-called 'early agreements', i.e. closing the procedure after the first reading.

Access to trialogues is thus privileged and restricted to the above actors who are the key figures throughout the process. Trialogue meetings are conducted behind closed doors. During first reading trialogues it is only the official Commission proposal (and accompanying public documents such as impact assessments etc.) that are publicly available. As the whole purpose of trialogues is secrecy, the debates are obviously not openly accessible and there are no verbatim records or minutes. Consequently, with regard to individual and ongoing dossiers, there is no information available from first reading trialogue meetings. In the second reading, the EP's opinion, the Council's Common Position and the Commission's comments thereon are available, but again there are no available documents from the trialogues themselves. This is also the case with conciliation trialogues.

The trialogues are not mentioned in the Treaties, but have obtained a more or less semi-formal status in the two aforementioned Joint Declarations between the Commission, the EP and the Council. In general it is stated that the:

> trilogue system has demonstrated its vitality and flexibility in increasing significantly the possibilities for agreement at first and second reading stages, as well as contributing to the preparation of the work of the Conciliation Committee. Such trilogues are usually conducted in an informal framework. They may[15] be held at all stages of the procedure and at different levels of representation, depending on the nature of the expected discussion. Each institution, in accordance with its own rules of procedure, will designate its participants for each meeting, define its mandate for the negotiations and inform the other institutions of arrangements for the meetings in good time.
>
> (European parliament, Council, Commission, 2007)

There is (arguably) a strong consensus among commentators that the swiftness and efficiency of the OLP in large part should be credited to this informal contact structure, which has eased and strengthened the cooperation climate between the EP and the Council (Shackleton and Raunio, 2003). It is particularly the latter that favours this practice, whereas the EP has expressed ambivalence as the process has become less transparent (only involving a handful of MEPs) and the debate has partly moved out of the formal channels and the Parliament itself. The character and functioning of the trialogues will be discussed extensively below.

The development of the OLP

Prior to Maastricht

The process leading up to the introduction and then the further development of the formerly called co-decision procedure can perhaps best be illustrated from the perspective of the EP. Hence, even if the Council is still the dominating house in EU decision-making, the introduction of the OLP has increased the EP's role in legislative tasks significantly. Prior to the Single European Act (SEA) the EP was only scarcely involved in legislation through a *simple consultation procedure* which, as the term indicates, means that the EP was only consulted and the Council had no obligation to follow the wishes of the Parliament. Despite this, 'by the mid-1970s, the Council consulted Parliament on virtually all legislative proposals referred to it except those of a purely technical or temporary nature' (Corbett *et al.*, 2007: 205). The introduction of direct elections to

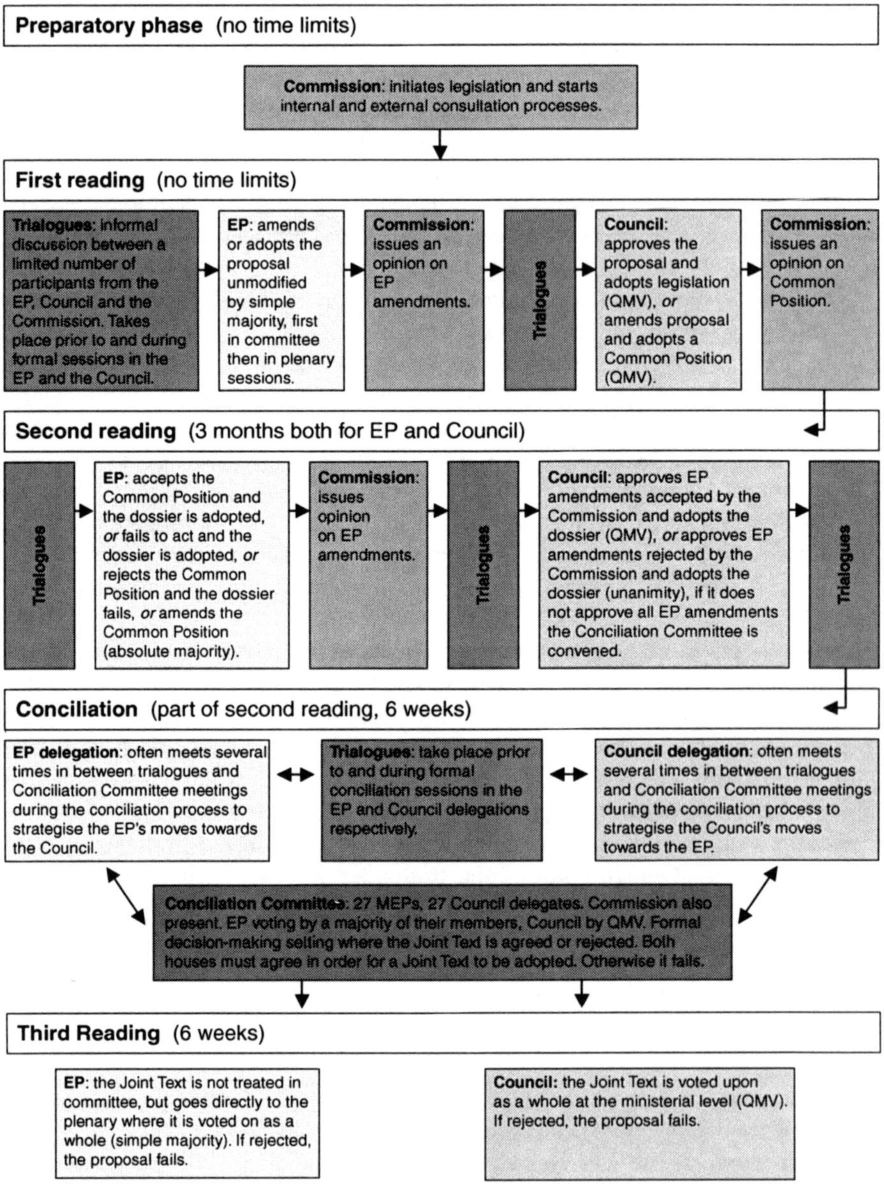

Figure 4.1 The ordinary legislative procedure.

the EP and the MEPs becoming full-timers from 1979 together with the important 'isoglucose ruling'[16] of the ECJ in 1980 signalled the more active role the EP was going to play in future EU decision-making. Now it was made clear that the Council was obliged to await the EP's opinion before it could adopt an act. However, the isoglucose ruling was more a power of delay than a veto and consequently more effective in situations where decisions required urgency. What is more, the Council could still disregard the EP's position if it did not agree (Hix, 2005: 78).

With the SEA, the powers of the EP grew as two additional decision-making procedures were introduced, namely the *assent*[17] and the *cooperation*[18] procedures. Now, both the number of policy areas where the EP was involved as well the powers of the EP were strengthened.[19] The cooperation procedure meant that the power of the EP was increased as it was now involved also in a second reading in decision-making processes. Procedurally this means that after the EP has produced its opinion, the Council must respond and agree on a Common Position prior to the second reading taking place: 'As a result, the habit of two readings gave the impression of a classical bicameral legislative procedure at European level, and helped pave the way towards full co-decision' (Corbett *et al.*, 2007: 214).

Maastricht – co-decision I[20]

The co-decision procedure was introduced with the Maastricht Treaty in order to improve the Union's democratic legitimacy through the strengthening of the EP's position and powers (Farrell and Héritier, 2003b: 6).[21] The EP was going to be a more equal partner to the Council in community legislation. As mentioned above, the rules regulating co-decision introduced the formal requirement that a proposal must be accepted by both houses in order to be adopted. This formal equality was symbolised by the establishment of the Conciliation Committee. Under the Maastricht provisions, recourse to the Conciliation Committee was, however, not necessarily the final step. If conciliation negotiations failed, the Council could re-introduce its Common Position and adopt the act unilaterally if the EP failed to drum up an absolute majority among its members to support its rejection within a six-week deadline. The Council only tried this option once in the 1994 Voice Telephony case where the EP managed to gather an absolute majority and voted the act down. The Council obviously found this strategy too costly and refrained from using this right in subsequent procedures (Corbett *et al.*, 2007: 215–6; see also Burns, 2006: 235–6).[22] Another provision that made the EP inferior to the Council was the former's obligation before the second reading to announce its 'intention to reject' the Council's Common Position 'and then confirm this in a new vote following an option conciliation meeting with the Council' (Corbett *et al.*, 2007: 215).[23] This provision was deleted in the Amsterdam Treaty.

Even if not as accentuated as the well-known *budgetary* quarrels in the 1970s–1980s (Hayes-Renshaw and Wallace, 2006: 215), the beginning of the Maastricht period was characterised by confrontation and lack of cooperation between the EP and the Council, as the legislative procedures hitherto had not warranted as close cooperation and interdependence as was required in co-decision. The institutions were thus not accustomed to working together on a more equal footing. The weaker EP struggled to win influence whereas the more powerful Council tried to minimise the former's new-won formal clout. The Council had no tradition for paying attention to or taking seriously the views of the EP. In these initial years of the co-decision procedure the EP, on its side, demonstrated a remarkable ability to act collectively and thus won long-term victories and more power, for example, through the aforementioned Voice Telephony case in 1994 (see e.g. Shackleton and Raunio, 2003; Burns, 2006).

Over time, the EP and the Council have developed a well-known and quite efficient arrangement of informal trialogue contacts and meetings where disagreements can be discussed and aired prior to formal discussions. The 'trialogues' or 'trilogies' were originally introduced in relation to conciliation meetings and the very first trialogue was held during the SOCRATES report in 1994 on the dispute over comitology and the budget (Benedetto, 2005: 70; Shackleton and Raunio, 2003: 177).[24] In the second half of 1995, under the Spanish Presidency, the EP and the Council agreed to hold regular trialogue meetings prior to the Conciliation Committee (Neuhold, 2001: 5). The reason was that the Treaty was, and still is, silent on what should happen between the Council's rejection in the second reading and what should occur prior to convening the Conciliation Committee. In addition to making the formal conciliation meetings much more prepared, the conciliation trialogues grew out of a frustration over conciliation meetings with over 100 people present and the lack of productive dialogue and negotiations between the EP and the Council (Shackleton, 2000: 334). Moreover, the formal conciliation meetings could be very time-consuming and difficult as the dossiers and hence the discussions were often of a very technical nature.[25] The trialogues prepare the conciliation meetings and function as an arena for sounding out differences and trying to find common ground. No formal decisions are taken here, as the participants must refer back the possible solution to their respective conciliation delegations for potential approval (Shackleton and Raunio, 2003: 177).

In sum, trialogues are not described in the Treaties, but now play a crucial role in the workings of the OLP. Their normative status is, however, highly ambivalent. I will discuss the trialogues thoroughly below, but for now it suffices to say that the trialogues represent an important feature in the development of a 'new legislative culture' between the EP and the Council and that under co-decision I, trialogues were first and foremost confined to the stage prior to and during conciliation.

Amsterdam – co-decision II

So far the introduction of the Amsterdam Treaty has represented the most important change to the co-decision procedure. First, the number of policy areas subjected to co-decision increased considerably, from 15 to 40[26] – spread over 31 articles (Shackleton, 2000: 326). Second, the provision whereby the EP had to announce its 'intention to reject' a Council Common Position was repealed:

> If the EP now votes to reject the common position, the legislation fails. This reform largely corresponds to the practice of co-decision: in the

> intergovernmental conference (IGC) negotiations, the Member States, the Parliament and the Commission all declared this part of the procedure to be moribund.
>
> (Maurer, 2003: 230)

Third, Amsterdam also removed the possibility of the Council re-introducing its Common Position after the second reading thus depriving the Council of its relative power to impose decision-making upon the EP unless the latter was able to mobilise an absolute majority and veto the proposal. Since Amsterdam, the future destiny of a policy proposal rests with the outcome of the Conciliation Committee – if no agreement is reached in conciliation the act fails and the procedure is closed.[27] This puts the Council and the EP on a more equal footing in legislation under co-decision (Shackleton, 2000; Benedetto, 2005: 69, Maurer, 2003: 230; Garman and Hilditch, 1998: 276). 'As a result, more than one observer has called the Parliament the big winner of the Amsterdam Treaty negotiations' (Shackleton, 2000: 326). Fourth, Amsterdam reduced the time limits for the duration of the co-decision procedure after the adoption of the Council's Common Position from 16 to 14 months (Maurer, 2003: 239).[28]

Finally, Amsterdam opened the way for the so-called 'fast-track procedure' allowing the legislative process to close after the first reading if both the EP and Council agree. This also resulted in the aforementioned Joint Declaration between the EP, the Commission and the Council on the practical arrangements for the co-decision procedure.[29] Here, the institutions committed themselves to cooperate in an amicable manner and, in short, we may arguably say that the document semi-formalised the usage of trialogues. Moreover, this was also followed up by the Council – which is the institution that has had to change its working methods the most. In a report from the Presidency and the Council Secretariat where closer and more accommodating cooperation with the EP was underlined, the trialogues were mentioned as a means that had made co-decision more effective, i.e. reduced the number of conciliations and increased closure at first reading.[30] Whereas the Maastricht period and co-decision I can be seen as the trial and error period, where the institutions tried to come to terms with the new legislative arrangements (i.e. the Council accepting and taking seriously the EP as a co-legislature), the Amsterdam period and co-decision II solidified the informal cooperation arrangements that developed under co-decision I. From Amsterdam onwards, the aim has been to reduce the number of dossiers going to conciliation and to try to reach agreement as early as possible, hence the need for more continuous cooperation and contact on an informal basis. Partly for this reason, Hix (2005: 77–8) argues that 'by the mid-1990s the EU legislative system had developed into something much closer to a traditional bicameral model'.

Nice

The Nice Treaty did not represent big changes with regard to the structure of the co-decision procedure and only increased the number of policy areas 'by five, immediately after ratification, and by five more, at a later date' (Shackleton and Raunio, 2003: 186). The co-decision procedure then covered over 40 areas of Community action. With regard to co-decision, the big question during the IGC was whether it should concur with all areas where decisions are taken by qualified majority vote (QMV) in the Council, as the EP fought for. This was, however, only partially implemented as 'among the new cases under QMV, three legislative ones remained outside the co-decision procedure: financial regulations, internal measures for the implementation of co-operation agreements, as well as the Structural Funds and the Cohesion Fund' (Rittberger, 2007: 135, see also Nugent, 2003: 88). The new policy areas subjected to co-decision after Nice were measures to combat discrimination, judicial cooperation in civil matters (excluding family law), specific industrial support measures, economic and social cohesion actions (outside the Structural Funds), the statute for European political parties and measures relating to visas, asylum and immigration.

Lisbon: from co-decision to the 'ordinary legislative procedure'

Even if the French and Dutch referenda formally put an end to the Constitutional Treaty, the majority of its content survived the 2007 IGC and was adopted at Lisbon. After two hurdles of popular referenda in Ireland, the Lisbon Treaty was finally ratified and came into force on 1 December 2009. One of the main tasks for the Convention on the Future of Europe was to simplify the legislative procedures in order to make the EU system more transparent and comprehensible. The Constitutional Treaty proposed to make the co-decision procedure the 'ordinary legislative procedure'[31] and to more than double the number of Community areas subjected to it, i.e. the majority of EU legislation. This is also the case with the Lisbon Treaty.[32] As in the Constitutional Treaty, the pillar structure is abolished in the Lisbon Treaty, thus paving the way for more EP involvement. Co-decision or the 'ordinary legislative procedure' now applies to important policy areas such as agriculture, fisheries and Structural Funds as well as the whole former third-pillar area of justice and police cooperation. Other decision-making procedures such as consultation and assent are brought together under 'special legislative procedures'. To make co-decision the default procedure in policy areas where the Council decides by QMV has been a principle advocated by the EP for a long time and the Constitutional Treaty seems 'finally to establish this link' (Hayes-Renshaw, 2006: 216). The Lisbon Treaty also adopted the clause providing for a more permanent leadership in the Council. However, the President of the European Council has not significantly affected the current set-up of co-decision-making, as the Council of Ministers will retain its rotating six months presidency when dealing with legislative acts.

As mentioned in the last chapter, another difference between Nice and Lisbon pertains to the new official role for national parliaments in the OLP through the early warning system[33] and the orange card procedure – the latter of which only applies when decisions are taken through the OLP.[34] Each national parliament[35] has two votes and if, within eight weeks, a simple majority (currently 28) of parliaments vote against the Commission's proposal, the latter is obliged to review the proposal and subsequently decide whether to withdraw, amend or maintain the proposal. If the Commission maintains its proposal it must justify its position through a reasoned opinion. Before the end of first reading and against the background of the national parliaments' and the Commission's reasoned opinions, the EP and the Council must consider the proposal against the subsidiarity principle. If both the Council (by a majority of 55 per cent of the members) and the EP (by a majority of the votes cast) find that there has been a breach of the subsidiarity principle, the proposal is abandoned. Hence, it is the rejection of the EP and the Council that ultimately stop a proposal, even if the national parliaments (through their reasoned opinions) are instrumental in starting the process. On the other hand and contrary to the national parliaments, the EP and the Council may decide that they want to proceed with the Commission's proposal whereby the national parliaments have no further 'sanctioning' possibilities. As we can see, the orange card represents a right to be heard, but is at no stage a red card to veto Commission proposals as the EP and the Council have. In this sense, the orange (and yellow) card procedure has strengthened the national parliaments and given them a formal role in the EU's ordinary decision-making. Despite this formal empowerment of the national parliaments, the relative impact and de facto influence of this right on EU decision-making under the OLP remains to be seen. As mentioned in the previous chapter, no orange card procedure has so far been initiated. According to Buzogány and Stuchlik (2011:5), this is 'in line with many expectations, very little has changed in the (admittedly relatively short) time period "after Lisbon".'

The OLP in numbers: efficiency and effectiveness

The acts coming out of co-decision-making are either regulations or directives and decisions. The number of concluded OLP acts has grown steadily from Maastricht until today. In the first co-decision period between Maastricht and Amsterdam there were 165 concluded dossiers. Within the fifth parliamentary term (1999–2004), the total number of concluded co-decision files was 403.[36] In the sixth parliamentary term (2004–2009), the number of adopted dossiers increased to 454 (European Parliament, 2009b: 6). For the current seventh parliamentary period, the EP reports that from 14 July 2009 till 14 March 2012, a total number of 195 dossiers have been adopted.[37] As can be seen from Table 4.1, the consultation procedure used to be the most often used legislative procedure in the EU. However, the number of acts adopted under this procedure has decreased substantially whereas the number of OLP acts has increased in correlation with the latter gaining more importance on the legislative arena.

Table 4.1 Adopted legislative acts 2005–2010[1]

Year	*Consultation*		*Cooperation*[2]		*Co-decision/OLP*		*Assent*
	1st reading		*2nd reading*		*3rd reading*		*Total co-decision*
2005	113	2	90	32	0	122	24
2006	158	0	74	34	10	118	13
2007	157	0	119	22	6	147	9
2008	110	0	140	29	1	170	10
2009	89	0	75	22	8	105	6
2010	38	Na[3]	80	14	0	94	23

Notes
1 All numbers taken from the General Reports on the Activities of the European Union from the respective years, see European Commission (2006, 2007, 2008a, 2009, 2010, 2011).
2 The cooperation has two readings, but since its numbers are so low I only refer to cooperation procedure acts in total.
3 With the Lisbon Treaty, the consultation procedure has been repealed.

Length of procedures

Many had predicted that greater involvement of the EP would mean slowing down and making the legislative process less efficient.[38] Studies show that this was not the case prior to the enlargement in the EU15, rather to the contrary: 'A more detailed analysis indicates that the average duration of co-decision fell from 769 days (for the period November 1993 – December 1994) to 409 days for proposals made between July 1999 and July 2002' (Maurer, 2003: 239, see also Hayes-Renshaw and Wallace, 2006: 212; Hix, 2005: 79).[39] However, in a recent study by Settembri (2007: 29) where he takes heed of the impact that the 2004 enlargement has had on EU decision-making, he found that with regard to the length of the decision-making process, his data also shows that the average number of days to complete an OLP file has now gone up to the same level as in the 1993–1994 period. More specifically, prior to enlargement he found that an average of 600 days was taken to close an OLP dossier, whereas after enlargement this number has risen to about 730 (ibid).[40]

Moreover and according to the EP (2009b: 12):

> There is also a clear trend for 1st and 2nd reading agreements to take more time. The average time until adoption increased between the 5th and the 6th legislature and also during the 6th legislature. A study of the Commission services found that the average length of time for a 1st reading agreement was 15.7 months and 32 months for a 2nd reading agreement.

In their study, Toshkov and Rasmussen (2012) nuance this picture further as they found that (especially) on salient and important OLP proposals, the duration of first reading processes ending up as early agreements (i.e. closure after first reading) last longer than first readings which proceed into second and third readings.

Fast-track procedure – early agreement

Despite the fact that the duration (number of days) of the legislative process is now more or less on a par with the early Maastricht situation this could imply that the dossiers also go through all the three readings. This, however, is no longer the case and as can be seen from Table 4.1, the overwhelming majority of OLP acts are now adopted after the first reading. As we have seen, this type of 'fast-track legislation' or 'early agreement' was made possible with the coming into force of the Amsterdam Treaty.

During the pre-Amsterdam parliamentary period (1993–1999) when it was *not* possible to conclude an OLP act at the first reading, 60 per cent or 99 dossiers were closed at second reading and 38 per cent or 63 dossiers were agreed after conciliation (Hayes-Renshaw and Wallace, 2006: 213).[41] In the fifth parliamentary period 1999–2004, 28.5 per cent of all OLP dossiers have been closed after first reading (i.e. 115 dossiers) whereas 49.6 per cent (i.e. 200 dossiers) were closed at second reading and 21.4 per cent (i.e. 86 dossiers) were finalised at third reading (see Hayes-Renshaw and Wallace, 2006: 213).[42] For the sixth parliamentary period (2004–2009)[43] the pattern of reaching agreement after the first reading has increased significantly. Of the 122 dossiers adopted in 2005, 74 per cent were completed at first reading. From the beginning of the seventh parliamentary term till 14 March 2012, 151 dossiers were concluded at first reading, 16 dossiers at early second reading and 21 dossiers at second reading, whereas 7 dossiers went all the way through conciliation and the third reading.[44]

The emphasis on first reading also means that there is hardly any dossier that goes all the way through the third reading, which is consequently almost a rarity today. It is first and foremost the Council that has been eager to close agreements in the first reading, seeking to avoid conciliation as much as possible and to 'reach agreement as swiftly as possible whenever possible' (Hayes-Renshaw and Wallace, 2006: 212). This 'commitment' to reaching agreement at first reading is also followed up in the aforementioned Joint Declaration from 1999 and revised in 2007 between the Commission, the EP and the Council on the workings of the OLP.[45] In addition to dossiers closing after the first reading, early second reading agreements have also become quite common and were 'formalised' during the previous (sixth) parliamentary term, according to the EP (2009b: 12). An early second reading agreement is reached in trialogue prior to the official start of the second reading and indicates that the EP accepts the Council's Common Position as it stands.

Conclusion

The aim of this chapter has been to provide some background information on the rules and practices of co-decision-making ranging from a historical synopsis of major developments from Maastricht until today, to a presentation of the legal bases and the functioning of the procedure (both the formal (three readings) and informal (trialogues) aspects) and finally a more quantitative summary of the efficiency and effectiveness of the OLP. Altogether this provides a foundation for the evaluation starting with the first criterion of democratic deliberative meeting places in the next chapter.

Part II

Evaluation

5 A forum for the public?

Are there democratic deliberative meeting places?

Introduction

In order to be able to assume that the adoption of OLP dossiers has democratic qualities, the first criterion warrants the existence of democratic deliberative meeting places. As argued earlier, it is not necessary (or possible) for a procedure to display democratic deliberation at every stage and in every setting of the decision-making process. However, in order to claim that the OLP is at all capable of producing democratic outcomes, the perspective of deliberative democracy depends – at minimum – on the existence of at least one open setting where those decisions are tested and critically examined by popularly elected representatives in a manner that is publicly available and accessible. Since there are two decision-making institutions involved in the OLP, the requirement is that both legislative houses provide one such setting each.

As discussed in Chapter 3, the indicator for democratic deliberative meeting places is that there is at least one setting in the EP and one in the Council that fulfils the following requirements:

- The setting is dominated by (a representative selection of) elected representatives.
- Access to the documents used and discussed in the setting is readily available in electronic format; discussions are conducted in openness by securing (translated) verbatim records and preferably (live and stored) internet transmission of meetings; the (verbatim) records identify who voted and what (if any) votes were cast.
- There are rules for ensuring and inducing equality among the participants: Rules should be (formally) described in the Treaties and/or rules of procedures/inter-institutional agreements stating the purpose of the setting and confirming that elected representatives are the main actors. The rules must describe the competencies, rights and duties of the setting. There must be rules for when and how the meeting ends (e.g. voting) and a commitment to announce the meeting agenda to all involved actors at the same time. There must be rules, governing debate and argument, which respect the principle of equality among the participants (unless there are good reasons given for why this should not be so).
- Time for debate and judgement prior to decision-taking is allocated.

Now, it may very well be that further investigation will reveal that the setting is not very important for actual decision-making or that it is not the popularly elected representatives who are the real decision-makers. These are valid and important points and will be tested through the other criteria (especially through criteria 2 (inclusion) and 5 (decision-making capacity)). Initially, however, we leave such connections (or lack thereof) aside and concentrate on determining whether there are relevant OLP settings that come close to meeting the requirements for a democratic deliberative meeting place.

Moreover, in order to determine the question of democratic deliberative meeting places, we need to know which settings are relevant under the OLP and on that basis select those that are likely to pass the threshold. As we can see from the indicator, for a meeting place to be eligible for the democratic category it must contain elected participants. Hence, this will be the mechanism for selecting the settings that will be considered further.

In this regard we can exclude all the meeting places that occur prior to the Commission's official launch of an OLP dossier as neither the internal Commission meeting places (such as the College, Chefs de Cabinets meetings, Special Chefs meeting, the DGs or inter-service consultation, etc.) nor the externally oriented consultation settings (such as the expert committees, online consultations, the policy/dialogue forums, etc.) consist of popularly elected representatives. However, when we move to the official part of the OLP, the following settings clearly meet the initial threshold: (1) the EP committees, (2) the EP plenary sittings, (3) the meetings of the EP conciliation delegation and (4) the ministerial meetings in the Council. These settings are (if only indirectly so in the Council), dominated by popularly elected participants and they are composed in a representative manner according to the numerical strength of the political groups in the EP and the national make-up in the Council.

There is, however, also a setting that is more difficult to determine, namely, (5) the Conciliation Committee. The Conciliation Committee has a kind of 'in-between' status, as half of the meeting consists of a representative selection of MEPs, whereas the Council is, at best, only represented by the minister holding the Presidency. The Conciliation Committee will consequently be included for further examination.

If we look at membership, we could, in addition, mention the trialogues as they also contain elected representatives, that is, the handful of MEPs representing the Parliament. However, the composition is not representative and is biased especially towards the two largest political groups (EPP and S&D). In addition, the Council is usually only represented by the Coreper from the member state holding the Presidency, as well as a few of the most central officials from the working groups. The ministers are never present apart from the minister holding the Presidency who may participate from time to time. Based on the above, the trialogues are not eligible for further assessment.

Some readers will perhaps wonder why I have excluded the EP political groups as relevant meeting places given that they are also dominated by elected representatives. The reason for this is that these meetings obviously do not contain a representative selection of MEPs. In addition, they are not conducted in openness – there is little information either from the political groups' own websites or in the academic literature on what is happening in these meetings with regard to OLP dossiers.

For decision-making under the OLP, the Council working groups and Coreper play crucial roles. These settings are, however, excluded as they only contain non-elected officials. In addition, many OLP legal bases require the opinions of the European Social Committee and the Committee of the Regions respectively, but these settings also do not contain popularly elected representatives. As mentioned above, the national parliaments now have an official role in the OLP through the orange card procedure. Whereas national parliaments obviously fulfil the requirement of elected representatives, they are not included in this book as the orange card procedure still remains to be initiated.

The structure of the chapter follows the assessment of the five meeting places respectively: (1) EP committees, (2) EP plenary sessions, (3) EP conciliation delegation, (4) the ministerial level in the Council and (5) the Conciliation Committee.

EP committees[1]

Elected participants?

In the EP, the committees play by far the most important role in parliamentary work, as is also the case under the OLP. Hence it is first and foremost the members of the responsible committee(s) who are involved in the drafting process.[2] The composition of the committees is regulated in rule 186 of the EP's rules of procedure and reflects the balance between the political groups in plenary.[3] Committees have both *full members* and *substitutes*, which means that most MEPs serve as full members in two committees and as substitutes in a third, or are full members of only one committee and substitutes in two.[4] Substitutes have the right to speak in committees, even if the full members for whom they are substituting are present, but they only have the right to vote when a full member of their political group is absent. The allocation of committee memberships is decided by the political groups (which set the broader guidelines) and the individual MEPs (in expressing their personal interests) (Judge and Earnshaw, 2003: 186). The number of seats for each political group is allotted according to numerical strength.

Committee meetings are also attended by the Commission and the Council. They have speaking rights, but no voting rights. The Commission sends officials (the rapporteur/chef de file) from the DG responsible for drafting the proposal and/or someone higher up in the hierarchy, sometimes even the Director-General or the Commissioner themselves if the dossier is deemed particularly important (Corbett *et al.*, 2007: 143). The Council is represented by officials from the Secretariat, Coreper and/or Presidency, but also from the Council working parties.

In sum, EP committees are composed and dominated by a representative selection of MEPs, but are also open for other actors.

Access to documents and debates?

In its rules of procedure, the EP provides a list of all EP documents that are available through its register of documents.[5] With regard to committee work, the relevant documents are: draft agendas, minutes, working documents, notices to members, draft reports, amendments to draft reports, reports, draft opinions, amendments to draft opinions, opinions and opinions in letter form.[6] These documents can be found on the EP committees' own websites and translated versions in many (if not all – especially for the working documents) of the official EU languages are available. In contrast to plenary sessions, as it is not required to write full verbatim records of committee meetings, there are consequently only minutes from committee meetings (cf. rule 194, rules of procedure). When seen in isolation, the minutes themselves are consequently not satisfactory as the exact wording of every speaker's intervention is not documented, that is, we do not get to know who advocated what. The minutes contain information about who attended the meeting, who spoke (but not what they said), the purpose of meeting, the number of amendments (and which) were adopted and the votes cast (but not who voted in favour, against or abstained). In other words, to take stock of how discussions proceeded thus becomes difficult by consulting the minutes alone.

However, since July 2009 the EP now provides video streaming of committee meetings and this helps to compensate for the lack of verbatim records. Even if full verbatim records in combination with video streaming would have provided the best solution, committee debates are, since the beginning of the seventh parliamentary term – and in comparison to earlier parliamentary terms – easy to follow, and it is easy to trace what took place in the meetings.[8] What is more, amendment documents contain the name of the MEP making the proposal, the suggested wording of the new amendment and a justification for why the amendment should be adopted. In order to get a good overview of the impact of the amendment, the working document of the EP committee is organised in two columns with the Commission's version placed in the left hand and the MEP's suggested formulation in the right hand column. The words to be changed/deleted/inserted are in italics, which makes it easy to get an overview. The second reading follows the same procedure, only this time the columns contain the Council's Common Position and the EP' second reading version.

The EP is a highly multilingual setting and whereas full translation is always ensured in plenary sessions, the policy is more flexible at the committee stage. According to rule 146, an interpretation of meetings should be provided by request to the members and substitutes.[9]

Hence, and as we shall see, in contrast to the Council (which prefers to keep the preparatory processes almost confidential) it is possible to reconstruct what took place in EP committee meetings through the public availability of the other working documents. The lack of verbatim records, however, makes it more cumbersome since one has to gather all the documents in order to piece together what was decided. This is obviously a drawback if the aspiration is to ensure the conditions for deliberative democratic scrutiny and testing of arguments in a publicly accessible manner, but the televised committee sessions now contribute towards improving ordinary citizens' ability to get a more thorough overview of what took place in the meetings. In comparison, it should be mentioned that many national parliaments keep their committee sessions in camera (e.g. German Bundestag).

Rules ensuring and inducing equality among the participants?

The EP committee meetings are not as such mentioned in the Treaties (neither in Article 294 nor in Articles 223–234 TFEU), but they are described in great detail in the EP's rules of procedure.[10] Overall, the EP has by far the most regulated settings (and, as we shall see, the provisions on plenary sessions are even more comprehensive). As an illustration, whereas the Commission's and the Council's rules of procedure are between 8–30 pages each, the EP's rules of procedure document is almost 300 pages long!

In legislative procedures, the committees prepare the parliamentary report to be discussed and voted upon in plenary. There are currently (in the 2009–2014 Parliament) 20 standing parliamentary committees in the EP,[11] but they are not all equally involved in OLP dossiers. In the EP's Activity Report (EP, 2009b: 9) for the last parliamentary term it is estimated that three committees are responsible for half of the OLP dossiers.[12]

In the committee there are a few *key positions* that are particularly important, namely the rapporteur, the committee chairs, the vice-chairs, the political group coordinators and the shadow rapporteurs (Mamadouh and Raunio, 2003: 339; Neuhold, 2007: 14–15). These actors do not only play a crucial role in the legislative activities inside the Parliament, some of them are also important participants in the trialogues. For this reason Benedetto (2005) calls them 'legislative entrepreneurs'.

Traditionally, the *committee chair* has perhaps been the most important position of the four.[13] The chair represents and speaks on behalf of the committee in plenary discussions, in the regular meetings of the Conference of Committee Chairs, s/he is the committee's representative outside the Parliament and is also automatically a member of the Parliament's delegation to the Conciliation Committee. Inside the committee, the chair, assisted by three vice-chairs, sets the agenda, organises and leads the meetings and decides who gets speaking time (Whitaker, 2001: 78). Along with the rapporteur, the chair also takes part in the informal trialogues and is informally consulted by the Commission before the latter tables a policy proposal marking the opening of the first reading.

Formally, each committee elects a *chair and three vice-chairs* (who together form the committee bureau) once every two and a half years. Actually, the election of the chair and vice-chairs is not a proper election, but mainly reflects the

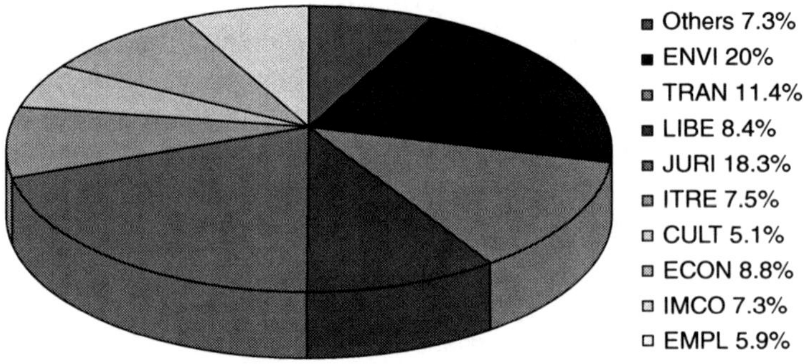

Figure 5.1 Diagram from EP (2009b: 8) indicating the distribution of OLP dossiers in the sixth parliamentary term by parliamentary committee.[1]

Note

1 It should also be noted that there is a different pattern with regard to when the various committees tend to reached agreement. The EP (2007a: 10) reports that it is first and foremost the Legal Affairs (JURI) and the Civil Liberties, Justice and Home Affairs (LIBE) Committees that can conclude dossiers in the first reading. The Transport and Tourism (TRAN) and Culture and Education (CULT) Committees, on the other hand, have a tendency to reach agreement in so-called early second reading, that is, after the EP, but before the Council has dealt with a dossier in the second reading. Finally, the Environment, Public Health and Food Safety (ENVI), the Industry, Research and Energy (ITRE) as well as the Internal Market and Consumer Protection (IMCO) Committees usually conclude their dossiers after a full second reading. According to the EP itself, the reason for these differences among the committees

> might be explained by the different 'cultures' of the major codecision committees, perhaps reflecting the smoothness of their working relationships with their counterparts in the Council formations and Commission Directorates-General. However, in part, such differences can only be explained by reference to the substantive points of disagreement between Parliament and Council on specific legislative dossiers.
>
> (EP, 2007a: 10)

size of, and agreement between, the political groups as well as the size of the national quota within the political group,[14] which is similar to the situation in most European national parliaments (Mamadouh and Raunio, 2003: 340, Judge and Earnshaw, 2003: 187). The political groups nominate their candidates and there are seldom competing candidates within the same committee (Corbett et al., 2007: 131). The allocation of positions is also determined by the composition in the previous parliamentary period.[15] The chair is allowed to take part in discussions and may vote, 'but without having a casting vote' (rule 195(4)).

Each political group has its own *coordinator* within the parliamentary committee. This person is chosen among its members.[16] The role of group coordinator is powerful in the sense that it is s/he who coordinates and sets the agenda of his/her political group in the respective committee, s/he distributes rapporteurships and decides who is allowed speaking time in plenary. Moreover, s/he also functions as a party whip, articulating the political group's position and making sure that the voting turnout is high (Whitaker, 2001: 68). Group coordinators from the various political groups within a parliamentary committee meet regularly to discuss pending issues, not least the distribution of rapporteur positions (Mamadouh and Raunio, 2003: 340).

According to rule 45(2) in the rules of procedure, the committee responsible for drafting the report shall appoint a *rapporteur* to be put in charge of the work. The rapporteur is thus, and in contrast to the committee chairs and the political group coordinators, appointed on a singular basis as upcoming legislation is presented and allocated to the relevant committee. He/she is chosen through a special weighting/auction system (not specified in the rules of procedure) reflecting the size of the political groups[17] – the guiding principle is thus proportionality. It is the chair and the political group coordinators who set the agenda and nominate rapporteurs for up-coming reports. Rapporteurs have usually been appointed once a dossier has been allocated to the responsible committee in plenary. However, rapporteurships have increasingly been distributed earlier, for example in relation to the presentation of the annual legislative programme (cf. rule 43(1)), in order to begin:

> preparatory work in advance of the formal presentation of a legislative proposal by the Commission, thus saving valuable time later on. It provides the possibility for rapporteurs to enter into talks with the Commission, before the latter finalises its proposal, as well as, in the case of co-decision, with the Council.
>
> (Corbett *et al.*, 2007: 141)

To be a rapporteur thus involves considerable legislative influence, as it puts the person in charge of drafting the report to be presented and defended to the EP in plenary sittings as well as in trialogues and the Conciliation Committee (Benedetto, 2005: 70, see also Yordanova, 2011). The rapporteur follows the dossier all the way through the legislative stages. Moreover, the report in itself plays a central role in the decision-making process and is a vital channel for influencing EU legislation, 'particularly where the Parliament can generate support in the Commission or where it enjoys a potential veto' – as under the OLP (Collins *et al.*, 1998: 6; see also Benedetto, 2005; Mamadouh and Raunio, 2003: 341). Apart from drafting the report, the task of the rapporteur is to initiate the discussions in the committee and to amend the report in accordance with the process of committee debates. The role is one of broker who builds up consensus among the members and defends the collective (or at least the majority) position even if this runs counter to his/her own position (Bowler and Farrell, 1995: 242). Besides drafting the report, the influence of the rapporteur is thus dependent on the ability to build consensus, first among committee members and second among the MEPs in plenary (Benedetto, 2005: 85).

It is, however, not only the responsible committee members who participate in the preparatory work. MEPs from other committees contribute and other committees also send their opinions.[18] When a case is deemed important, the political groups, especially the larger ones, often appoint so-called *shadow rapporteurs* to follow the work of the dossier closely. Shadow rapporteurs perform important functions in the sense that they provide well-prepared opposition (e.g. suggesting amendments) and mobilise their group both in plenary and committee sessions and sometimes also participate in trialogues. What is more:

> In some cases they (or Group coordinators) practically constitute informal sub-committees. A particularly striking example of this was the REACH chemical legislation, where the rapporteur and shadow rapporteurs as well as the draftsmen of the two opinion-giving committees met on an almost weekly basis over an extended period of time to identify the main area of agreement and possible solution for the key problems at stake.
>
> (Corbett *et al.*, 2007: 141)

To sum up, even if all committee members have an equal status and have the right to suggest amendments, the above key positions make some actors more powerful than others. This, in principle, is not a problem as long as access to such positions is open for all and not constantly occupied by the same individuals. If all MEPs have a fair chance to take on the different positions, it can serve as a practical arrangement dividing committee work among its members. As we have seen above, all the key positions are open to all committee members and are in this regard not contrary to the principle of equality. However, they are more often occupied by the MEPs from the largest political groups (EPP, S&D, ALDE). In fact, it is the political groups that 'run' the EP in the sense that all key positions are distributed according to group membership and group strength. Hence the three largest political groups[19] are in charge of how positions are allocated in the EP to the disadvantage of not only non-attached MEPs, but also the MEPs from smaller political groups.[20] On the other hand, taking on rapporteurships is quite a demanding task and even if all MEPs obtain assistance from the EP Secretariat when preparing the report, it is time-consuming and resource-demanding and thus easier for MEPs with a sizeable and organised political group to back them. Partly for such reasons, it is not entirely straight-forward to determine the level to which the rules work to the disadvantage of non-attached MEPs and MEPs from smaller parties (see Corbett *et al.*, 2007: 98).

The Council is famous for its consensual way of reaching agreements even in policy areas where QMV is the rule (Hayes-Renshaw and Wallace, 2006). Interestingly Settembri (2006) investigated voting patterns in EP committees during the 1999–2004 parliamentary period. He looked only at legislative reports (i.e. including OLP procedures) and found that the average majority of MEPs endorsing committee reports was as high as 93.7 per cent.[21] The OLP was no exception, with an average majority above 90 per cent on all reports adopted. Moreover:

> the size of the consensual bloc does not decrease (or increase) as co-decision procedure moves from the non-binding first reading to a final vote on the compromise reached in the conciliation committee. The size of this majority does not vary significantly over time or across parliamentary committee either.
>
> (Settembri, 2006: section 2.3)

There is one important difference between the first and the second reading with regard to who can table amendments. Whereas all MEPs are allowed to table amendments in the first reading, only full members and permanent substitutes in the responsible committee have this right in the second reading (cf. rule 63). In the third reading, the dossier goes directly to the plenary. The committee session in the second reading ends by putting the amended proposal to a vote and it is adopted if there is a majority of the votes cast.[22] The reasoning behind limiting the number of MEPs who are allowed to suggest amendments in the second reading is arguably an efficiency choice. However, it may be legitimated on the basis that the responsible committee is composed of a representative selection of MEPs. Hence it can be argued that the rules still ensure a representative selection of opinions and positions even if not all MEPs individually have a say in the second reading.

It is the chair or the EP President who can convene committee meetings (rule 193(1)). Even if the seating arrangement is formal, with the chair and vice-chairs sitting on a podium vis-à-vis the other members (cf. Corbett *et al.*, 2007: 134), the atmosphere in committee meetings is less formal than in plenary sessions. This is the case both with regard to translation and voting.[23] In committee meetings, non-translated or oral amendments can be accepted especially if these are presented by the rapporteur, the chair or the group coordinators in an effort to find compromise solutions (Corbett *et al.*, 2007: 144). Voting procedures in committee are also less formal and (unfortunately) often to the detriment of transparency. Whereas it is technically possible to conduct roll-call votes or calling for a quorum, this is hardly ever used. Instead, voting is usually conducted through the showing of hands or, in cases where the chair of the meeting is uncertain about the result, by electronic vote.[24] The report must contain the suggested amendments with short justifications 'if appropriate', and a draft legislative resolution, possibly also accompanied by an 'explanatory statement including a financial statement which establishes the magnitude of any financial impact of the report and its compatibility with the multiannual financial framework' (rule 45, EP rules of procedure). It is also possible to attach a short statement if there is a minority position (ibid, rule 52).

Time for debate prior to decision-taking?

Time for discussion is not specifically mentioned in the EP's rules of procedure, but is implicit in the very purpose of committee meetings.[25] However, the proceedings of the meetings are described in the academic literature on the EP. On average, busy committees meet 'at least twice a month for two, three or four half days on each occasion' (Corbett *et al.*, 2007: 134). In the first reading, the work of the committee is focused on responding to the Commission's initial proposal. More specifically, the opening of a dossier is devoted first to a presentation of its content by the rapporteur, subsequently followed by a discussion where committee members as well as the Commission and the Council can take part (Corbett *et al.*, 2007: 143).[26] This is an open discussion which seems to be oriented towards mapping the nature of the case and sounding out problematic issues: 'At the end of the initial discussion it is often clearer as to whether the issue is controversial within the committee or not' (Corbett *et al.*, 2007: 144). After this initial discussion, the rapporteur produces a text that is sent (after it is translated into all official working languages) to the MEPs and subsequently presented in committee where:

> There follows a fresh discussion, with committee members and the Commission commenting on the rapporteur's text. A deadline for amendments may then be set. The rapporteur can also decide to make modifications to his or her text or even to draft a new if there are a large number of amendments.
>
> (Corbett *et al.*, 2007: 144)

The EP itself also reports that it is quite common to conduct hearings on 'controversial or "technical" dossiers ... with experts or to commission impact assessments'.[27] In other words, to quite a large extent committee work is, as is the whole legislative process, a down to earth and very concrete affair as it amounts to building up support for the proposed text or deleting, inserting or reformulating amendments to the Commission proposal. Neunreither (2002: 53) even suggests that the focus on amendments often hinders more principled and overarching debates as it skews the attention to voting on single amendments rather than on the dossier as a whole: 'There are serious complaints that quarrels about amendments have to a great extent replaced a truly parliamentary debate. Some committees occasionally have to deal with 300–400 amendments on one proposal, a nightmare in itself.'

Towards the end of the process, the committee first votes separately on each amendment and finally on the whole report. A simple majority is needed for the whole report (Mamadouh and Raunio, 2003: 342), but before the vote is cast, it is customary to ask both the Commission for its position and the Council for comments. The intention is to sound out the reactions before the report is sent to the EP plenary, but also to 'put pressure on the Commission' (Corbett *et al.*, 2007: 144). This invitation notwithstanding, it is very rare that the Commission or the Council take the opportunity to respond – the Commission because it has 'usually made its approach clear in the preceding committee discussions' and the Council because 'it normally has no position at that stage' (Corbett *et al.*, 2007: 144). In the end, a legislative report from the committee should comprise suggested amendments and justifications for these, a draft resolution and an explanatory statement (including a statement on potential financial impact) (cf. rule 45). In other words, the EP committee meetings are organised to ensure debate prior to decision-making.

In sum, the set-up and practice in EP committees complies with the institutional conditions for a democratic deliberative meeting place. The committees are composed of a representative selection of MEPs and the access to documents and debates thus appears to be satisfactory even if there is a lack of full verbatim records and roll-call votes are missing. Moreover, the rules and practices for committee work are (presumably) organised according to the numerical strength of the ideological groupings. In this way we may say that the principle of equality is operationalised to reflect the proportionality principle underpinning the outcomes of EP elections and can consequently be justified from a democratic perspective. What is more, the organisation of EP committees also allocates time for debate although the nature of OLP legislation, where the focus is on tabling and voting on amendments, seems to hinder a more principled and overarching debate about the political impact of the dossier as a whole.

EP plenary sessions

Elected participants?

After the committee has finished its report, it is presented and discussed in EP plenary sessions where all MEPs are included. Plenary sessions are also open to other types of actors even if some are more privileged than others. The Commission[28] and the Council have special seats in the Parliament and there are also seats available for special guests such as heads of state and government. There is of course also space allocated for the translators. Whereas other actors may be invited to speak in Parliament, it is only the MEPs who have the right to vote.

Access to documents and debates?

Plenary meetings are conducted in full openness with both verbatim records[29] and video streaming of the sessions in all the official languages available (rule 182).[30] The public, together with lobbyists, the press, diplomatic services etc. cannot enter the room as such, but can follow the discussions from the gallery if they have a valid admission card (rule 145). When the EP treats a dossier in plenary it works with two types of documents, the Commission's proposal (with potential amendments) and the Parliament's draft legislative resolution. The suggested amendments are presented, discussed and voted upon on an individual basis before the entire document is put to a final vote. The voting procedure also follows this sequencing, that is, first the plenary votes on the Commission proposal with suggested amendments (both individual and to the whole text). Then the Commission is invited to react to the vote. Further:

> If the Commission gives an unsatisfactory response, or is not in a position to react immediately, the vote on the draft resolution may then be postponed for a period [of max two months] sufficient to allow the responsible committee time to examine the situation.
>
> (Corbett *et al.*, 2007: 177)

The procedure is not finalised until the EP has also voted on the draft resolution. However, the sequencing of discussion and voting is separated so that a report with or without amendments is not followed directly by a vote. EP plenary sessions have special voting times, usually scheduled to noon on Tuesdays, Wednesdays and Thursdays during the week of plenary sessions (see Corbett *et al.*, 2007:174–6).

After a report has been adopted with or without changes, it is accordingly amended by the rapporteur and his/her staff and subsequently sent to the Official Journal for official publication as well as to the Commission and the Council. The text of the EP's official position is not as pedagogical as the committee report with two columns providing the possibility of comparing the Commission's and the EP's formulations all the way through, but also, the official text is quite easy to get an overview of in the sense that the EP's proposed amendments are clearly highlighted in italics.

Rules ensuring and inducing equality among participants?

The provisions on plenary sessions are even more extensive than those organising the parliamentary committees.[31] In the first reading there are no time limits, but legislative dossiers are given priority when the agendas for plenary sessions are drawn up (Corbett *et al.*, 2007: 172). After more informal consultation among the committee chairs the agenda is adopted by the Conference of Presidents at their meeting, which takes place on Thursdays, before the monthly Strasbourg plenary sessions. There is a long description of the rules for how to adopt the draft and the final agenda (see rules 137–40). In the end, the final agenda is 'distributed to all MEPs and voted on by Parliament at the beginning of the part-session on the following Monday. The draft concerns the whole monthly "part-session" (i.e. includes any subsequent Brussels sittings that month)' (Corbett *et al.*, 2007: 172). Amendments to the agenda can be announced to the EP President up until one hour before the sessions start if they have the support of one political group, a committee, or 40 members. The 'President may give the floor to the mover [the MEP suggesting the amendment], one speaker in favour and one speaker against, in each case for not more than one minute' (rule 140). Moreover, the President ends the plenary session by announcing the time, date and agenda for the next session (ibid). In sum, we may say that the rules to ensure that the agenda is made known to every MEP are complied with.

According to rule 148, the relevant documents for the plenary session shall be distributed to the MEPs beforehand.[32] The legislative focus is now on responding to the committee report and it is possible to table amendments if these are suggested by the responsible committee, one of the political groups or at least 40 MEPs (rule 156).[33] In plenary, it is the EP president who decides which amendments are admissible or not. Amendments shall be presented in writing and before being voted upon, they shall be translated into all the official languages (ibid). In the first reading, there are no restrictions with regard to who can table amendments, whereas the second reading is in this regard more restrictive. In order to narrow down the process only amendments that have a foundation in already tabled suggestions or that seek to establish a compromise between the

EP and Council may be considered.[34] The MEPs vote on every single amendment which requires absolute majority in order to pass. Only a simple majority is needed in order to approve the Council's Common Position without amendments. In the end the whole proposal is put to a final vote. In contrast to the first reading where Parliament divides the voting sequence between the Commission proposal with the EP's suggested amendments and the responsible EP committee draft resolution, in the second reading there is only voting on the legislative text – i.e. the draft resolution is excluded – (Corbett *et al.*, 2007: 177). Whereas the first reading only requires that a simple majority of MEPs support a proposal in order to formally adopt it, the second reading requires absolute majority.

In the third reading, the dossier is taken directly to plenary session without any further treatment in committee. According to rule 69, the Joint Text shall be accompanied by a report and a statement, which shall be presented by the chairperson of the Conciliation Committee or another designated member of the delegation, usually the rapporteur who is the author of the report and statement. At this stage there is no room for opening up a debate on individual amendments and the task of the plenary setting is to vote on the Joint Text as a whole within six (possibly eight) weeks after agreement in the Conciliation Committee. A majority of the MEPs must support it in order to approve the act. If, however, the Joint Text fails to obtain a majority in its favour, the chairperson or another MEP from the delegation is obliged to make a statement, which is then followed by a debate.[35]

The latest revisions of the rules of procedure have resulted in a situation where non-attached MEPs now have less influence over when and which questions can be posed to the Council and the Commission; they are less likely to be able to table amendments, they now have less speaking time and they have less influence over the nominations of officers (Settembri, 2004: 152–3). Moreover, for these MEPs the situation has become worse than before as they have usually focused on making an impact in plenary settings rather than in committee 'through participation in debates, points of order or even demonstrations and challenges to the authority of the presiding offices. Their protests are rarer in committee and non-attached members are seldom allocated the more significant rapporteurships' (Corbett *et al.*, 2007: 98, see also Settembri, 2004). Hence, the non-attached members do not have anyone to represent and are therefore excluded from the possibility of participating. The same logic applies to the Conference of Presidents where one seat is reserved for non-attached MEPs, but unlike the other MEPs, they do not have the right to vote (rule 24, see also Settembri, 2004: 156). The bottom line is that a non-attached member will most likely never obtain one of the above-mentioned key positions. We should, however, remember that there were only 30 non-attached MEPs in both the sixth and the seventh parliamentary terms. The situation of non-attached members has been hotly debated in the Parliament following the establishment of the TDI group – a group of non-attached MEPs forming a purely and explicitly apolitical, technical group in order to obtain the benefits of belonging to a political group (more financial resources, administrative staff, more speaking time etc.).

This group (after consideration by both the EP and the ECJ) was eventually dissolved and it is now a clear requirement for the formation of political groups that they must have shared political affinities.[36] In sum, the rules underpinning the workings of the plenary sessions are seemingly conducive towards ensuring equality among the MEPs even if the non-attached MEPs come out disadvantageously when it comes to time and resources and to filling key positions which would arguably make them more influential.

Time for debate prior to decision-taking?

The EP plenary sessions are held for one week once a month. The President of the EP draws up a proposal for the allocation of speaking time for up-coming plenary sessions. The proposal is discussed and authorised by the political groups.[37] In addition to MEPs, Commission and Council representatives are also allowed to speak in EP plenary sessions. A dossier is usually first introduced by the Commission and subsequently commented on by the rapporteur who is presenting the committee report. In addition, other committees may present their opinions. Then follows a debate where the largest political group starts, followed by the next biggest group etc. and ending with the non-attached members (rule 149). In order to keep to the timetable, plenary debates have strict speaking rules, which are overseen by the EP President who chairs the meetings (ibid). When a political group or other actors have used their allocated time, the chair asks the current speaker to stop and if this is not complied with the microphone is simply switched off (Corbett *et al.*, 2007: 174). Apart from the President, other MEPs are (in principle) not allowed to interrupt a MEP when s/he is speaking (rule 149).[38] What is more, a debate cannot be closed before the list of speakers has been exhausted unless it has been voted on (rule 176).

Debates on legislative dossiers end with the Commission indicating its position, reactions to the discussion and the amendments suggested in the committee report. Unlike other speakers, the Commission is, according to Article 230 TFEU, under no time pressure in this situation and has ‘an unlimited right to intervene’ (Corbett *et al.*, 2007: 174). The Commission thus often ‘eats’ into the time allocated to others. In addition, the Council will often reply towards the end of such a debate, but it has no ‘carte blanche’ with regard to speaking time. For MEPs:

> Time for debates is thus strictly limited, with a corresponding lack of scope for spontaneity, or cut-and-thrust exchanges between individual members. This is reinforced by a tendency for some members to read out their speeches, repeating set positions that have often already been aired in committee.
>
> (Corbett *et al.*, 2007: 174)[39]

The multilingual character of the EP obviously also hinders spontaneity, at least for some MEPs (Corbett *et al.*, 2007: 180). Still, even if its debates may not attract as much attention as those in national parliaments, 'it has developed a method to enhance the role of its members in actually shaping policy outcomes rather than being a rubber stamp or a simple forum' (Corbett *et al.*, 2007: 183).

This last point notwithstanding, the number of participants present in plenary sessions, the speaking rules and the time limitation are all factors that seemingly challenge the possibility for thorough argumentative exchanges between the MEPs. However, we can see the organisation of the EP as sequential starting with the committee and ending or finalising the process in plenary settings, i.e. as a division of labour between deliberation and decision-making:

> The committee is engaged in drafting (and redrafting) proposals; in giving detailed consideration to proposals line-by-line; in fact-finding and in framing appropriate responses to the facts found. This is creative, cooperative work. The parent assembly, in contrast, is more in the business of disposing of proposals, affirmatively or negatively. Discourse in the former venue is aimed more at refining proposals. In the latter venue, it is aimed more at rationalizing acceptance or rejection.
>
> (Goodin, 2005: 188)

Based on the above, the quote from Goodin can also be seen to reflect the relationship between committees and plenary in the EP. Hence, in addition to the level of openness to which these meetings are subjected, the EP's heavy reliance on treatment in committee prior to more general discussion in plenary can be justified on the basis of the representativity of the composition of the committees. That is, 'the parent assembly delegates certain elements of its deliberative task to these other subgroups', i.e. committees, where it assumingly stands a greater chance of being subjected to deliberation (Goodin, 2005: 187).[40]

In sum, EP plenary sessions are carefully described and regulated through detailed provisions in the rules of procedure where the foremost objective seems to be ensuring a certain level of equality between the MEPs. The rules range from the convening of sittings to the proposal of amendments, speaking rules, access to documents, translation, voting procedures, etc. Hence also the set-up and practice of the EP plenary sessions complies with the necessary requirements for a deliberative democratic meeting place, even if the number of participants and the time constraints condition and restrict the type of debate that it is likely to expect. It is, however, due to the openness of debates and access to documents that it is possible for Union citizens, and the public in general, to follow the debates and to know the position and argumentation of individual MEPs.

EP delegation meetings

Elected participants?

For every dossier reaching conciliation, the EP appoints a 'new' delegation.[41] The EP delegation is based on the political composition[42] in Parliament and is assisted by the Conciliations and Codecision Secretariat ('CODE').[43] In the current seventh parliamentary term, and based on the outcome of the 2009 election, the distribution of seats is as follows: EPP 11 members, S&D (former PES) seven, ALDE three; the Greens and the ECR[44] have two members each, and the GUE/NGL and the EFD have one member each. Of these, one of the three EP Vice Presidents is always part of the delegation[45] as are the rapporteur and the chair of the responsible EP committee (the two latter as ex officio members). It should be noted that neither in the sixth nor the seventh parliamentary periods were representatives from the non-attached MEPs included.[46] The Vice President is the head of the delegation. The political groups may also appoint substitutes who can attend meetings, but who can only vote when the full member is absent (Corbett *et al.*, 2007: 223). The majority of the delegation members come from the EP committee responsible for the dossier. The names of the MEPs participating in the various conciliations can be found on the EP's internet site on conciliation.[47] As in all other EP meetings, the Commission is represented, but the Council is (obviously) excluded (EP, 2007b: 13, 16). The EP delegation thus consists of both permanent and temporary representatives. Since the composition of the delegation almost reflects the political landscape of the Parliament and all the delegates are elected politicians, it can be argued that the institutional set-up of the parliamentary delegation facilitates the conditions for including the views of affected parties. It should, however, be noted that MEPs from small parties, as well as non-attached members, appear less often than MEPs from the largest groups.

Access to documents and debates?

In order not to reveal the limits of flexibility and negotiation points to the Council, the meetings of the delegation are held behind closed doors (cf. rule 68, EP rules of procedure; Shackleton and Raunio, 2003: 178). The most interesting document at this stage is the 'joint four-column working document' which is confidential throughout the conciliation period, but released and made available on the EP's register of documents after the process is over (EP, 2007b: 21).[48] Other EP delegation and conciliation documents can be accessed at the EP's register, including working documents and records of attendance.[49] As in the EP committees (and unlike plenary sessions) the translation policy is more flexible in delegation meetings (cf. rule 146). As we can see, the EP delegation meetings do not comply with the requirements of access to documents and debates that are expected from democratic deliberative meeting places.

Rules ensuring and inducing equality among the participants?

The EP delegation meetings are not described as such in the Treaties, but are regulated in rule 68 of the rules of procedure and contain provisions on membership (and substitutes), voting rules (majority rule), who is chairing the meetings (the Vice President), that meetings are held in camera and that the delegation is accountable to the Parliament through reports and to the Conference of Presidents that decides the long-term guidelines of the conciliation delegation.[50] The meetings of the delegation are further described in the EP's Activity Reports on the OLP as well as in the Co-decision Conciliation Guide.[51]

Whereas all the delegation members meet in the Conciliation Committee proper, and are thus on an equal footing, only a selected number of MEPs participate in the trialogue meetings that take place in between meetings of the delegations and the Conciliation Committee. From the EP, it is the rapporteur, the chair and the relevant vice-president that are the so-called 'negotiating team'.[52] Hence by virtue of their participation in trialogues, they also become key actors in the delegation meetings. And even if the meetings are closed to the public, MEPs who are not participating in the conciliation delegation have – unlike the situation in trialogues – access to the meetings. Moreover, the rules were changed in 1998 in order to make:

> proceedings less confidential and to enable substitute members to enjoy greater participation in the conciliation committee. In general, the deliberations of the delegations are open to all those working inside the Parliament, and the conciliation secretariat makes a summary record of each delegation meeting.
>
> (Rasmussen, 2005: 1025)

The EP conciliation delegation meeting is not an official decision-making arena. Hence there are no formal provisions on voting or on how the delegation shall reach agreement, but according to the EP (2007a: 6) itself: 'The delegation aims to act by consensus and the key point is that the views of the individual political groups are expressed and that proper account is taken of these.' In order to ensure this an important task of the Vice President, who chairs the meetings, is to build up consensus and decide when there is agreement.

During conciliation, the institutions work on a common document, i.e. the aforementioned 'joint four-column working document'. It is the Secretariat who prepares the documents and sends them out to the MEPs. Prior to meetings, the note contains information on the aims of the upcoming meeting, the amendments, the state of the dialogue and negotiations with the Council, etc. After each meeting a summary report is drawn up and circulated to the members (EP, 2007b: 14).

The two last enlargements have created the problem of ensuring that a full translation of all meetings is available. However, the joint working document is available in the languages of all delegation members, but not always in all the languages of the substitutes (EP, 2007b: 14).

In sum, as with the EP committee and plenary meetings, it can also be argued that the delegation meetings are organised with a view to ensuring equality among the MEPs.

Time for debate prior to decision-taking?

For the EP delegation, the initiation of a conciliation procedure starts with a constituent meeting whose foremost purpose is to agree on a mandate for the conciliation trialogue team. According to the EP (2007b: 13), the meeting is described in the following manner: 'there is often a short exchange of views on the substance of the issues at stake, particularly when the initial reaction of the Council to Parliament's second reading amendments is known.' In addition, the Commission lays out its position on the EP's second reading amendments and potentially provides information about the Council's position.

The purpose of the delegation meetings is to respond to the proceedings in the trialogues, to discuss compromise texts and to provide the negotiating team (i.e. the trialogue participants) with further mandate and instructions on how to proceed in the next informal meeting. The delegation is kept continuously updated (in written form also) on the proceedings of the trialogues and on this basis discusses whether to accept or reject potential points of agreement. Moreover:

> prior to each delegation meeting, members receive a note about the aim of the meeting and the status of the negotiations plus other relevant documentation, in particular a revised version of the joint four column working document, where positions of the different EU organisations and possible compromises are presented.
>
> (Rasmussen and Shackleton, 2005: 14)

In this sense, we may say that delegation meetings have allocated time for discussion, but given the circumstances of conciliation, none of these discussions are conducted in openness.

In sum, the EP conciliation delegation meetings are not designed for ensuring publicly accessible discussions and testing of arguments. Rather, the intention of delegation meetings is to prepare the final negotiating rounds with the Council in the trialogues and subsequently in the Conciliation Committee. Hence, even if these meetings have certain democratic qualities pertaining to the inclusion of an almost representative selection of MEPs as well as access to meetings for those MEPs who are not members of the conciliation delegation, they are closed to the public. This does not mean that the meetings are unjustified in themselves, but it means that they cannot figure as representative of a democratic deliberative meeting place under co-decision-making.

The Council of Ministers

Elected participants?

A Council ministerial meeting includes all the ministers of the relevant configuration, assisted by their national delegation officials.[53] The Council Secretariat is present as it has the responsibility for preparing and coordinating meetings. Whereas officials may speak in these meetings, it is only the ministers themselves who have the right to vote. If one minister is absent and represented by the national permanent representative, s/he cannot vote on behalf of his/her minister, but this right can be entrusted to a minister from another member state (cf. Article 239, TFEU). Whereas the Commission does not have the right to vote, it participates almost as an equal partner or as a 28th member state in Council meetings and 'speaks, proposes and defends its positions, and is a full participant in negotiations within all areas where there is a clear Community competence' (Hayes-Renshaw and Wallace, 2006: 192, see also 199). In contrast, the MEPs do not have access to ministerial meetings (or any other relevant OLP settings in the Council such as working groups and Coreper meetings). Finally, interpreters attend the meetings, but they have no right to speak or vote. The ministers of the member states are the main actors in ministerial meetings and are the only ones with both speaking and voting rights. In addition, they have popular backing and thus the Council ministerial meeting can be said to be a *potential* deliberative democratic meeting place.

Access to documents and debates?

Over the last few years, Council ministerial meetings have become more open. In 2006 the Council decided to open up deliberations on OLP dossiers including the voting procedures if conducted.[54] This commitment to openness is further strengthened with the coming into force of the Lisbon Treaty, together with Article 16(8) TEU and Article 15 TFEU, where the Council is now obliged to discuss and vote on OLP dossiers in openness. In concrete terms, this means that from the Council's website it is possible to retrieve some of the relevant documents. These include; provisional agendas,[55] minutes[56] and voting charts.[57] In addition, the Council also offers monthly online summaries[58] of adopted acts and minutes of ministerial meetings. The minutes are, however, very schematic and not very informative for ordinary citizens (Hayes-Renshaw and Wallace, 2006: 54).[59] Council minutes covering legislative acts are accompanied by relevant statements[60] made by the member states or others, as well as the results and explanation of votes. Another source of online information is the press releases.[61] In sum, although there is some documentation of ministerial meetings, neither of the above-mentioned documents provides a record of the substantive content of dossier discussions during procedure.

Since July 2006, the Council decided to open up all deliberations on OLP dossiers including voting procedures if conducted (Council, 2006a). This is now reflected in Article 7 of the Council's rules of procedure:

> (1) The Council shall meet in public when it deliberates and votes on a draft legislative act. To that end, its agenda shall include a part entitled 'Legislative deliberations'. (2) Documents submitted to the Council shall be made public, and likewise those sections of the Council minutes which relate to

> that part of the agenda. (3) The opening to the public of Council meetings relating to the 'Legislative deliberations' part of its agenda shall be made through public transmission by audiovisual means, notably in an overflow room and through broadcasting in all official languages of the institutions of the European Union using video streaming. A recorded version shall remain available for at least one month on the Council's Internet site. The outcome of voting shall be indicated by visual means.... (4) The results of votes and explanations of votes by members of the Council or their representatives on the Conciliation Committee provided for under the ordinary legislative procedure, as well as the statements in the Council minutes and the items in those minutes relating to the Conciliation Committee meeting shall be made public.

By comparison, in the former version of the Council's rules of procedure the commitment to open deliberations was covered in Article 8 and was worded more strictly:

> (a) the presentation, if any, by the Commission of its legislative proposals and the ensuing deliberation in the Council shall be open to the public; (b) the vote on such legislative acts shall be open to the public, as well as the final Council deliberations leading to that vote and the explanations of voting accompanying it; (c) all other Council deliberations on such legislative acts shall be open to the public, unless, on a case by case basis, the Council or Coreper decides otherwise with regard to a given deliberation.

Presumably, the 2009 version of the rules of procedure represents an improvement from the 2006 version where, as we can see, the conditions of openness for OLP dossiers were more narrowly defined in the sense that open deliberations are now to be the rule throughout the ministerial meetings, and are not confined to the opening deliberations (when dossiers are presented) and the voting process. In addition, the Council now divides its agenda into two parts, a legislative part and a non-legislative one which allegedly also improves accessibility.

The sessions are in many cases televised and accessible from a video streaming service on the internet in all official languages (usually available in three or four languages and always in English and French). A recorded version of these public sessions must be stored and made available on the Council's internet site for at least one month. The General Secretariat is obliged to announce upcoming sessions in due time so that the citizens can orient themselves to the legislative work of the Council. Like the EP's online service, the Council's is also easy to manoeuvre one's way around. There is a table of contents attached so that viewers can skip and scroll down to the topics they are most interested in and which member states the speakers are from is normally indicated. However, and more importantly, when doing a few blind tests cross-checking provisional agendas (indicating that OLP issues were supposed to be discussed) with the stored live footage, only taped press conferences were available.

In sum, despite some shortcomings, the Council ministerial meetings 'score' better than working parties and Coreper with regard to openness.[62] Access to documents is quite good, so is openness of debates/votes with a combination – at least in many instances – of both minutes (not verbatim records[63]) and live video streaming. It should, however, be noted that since the Council is legally a unitary body it is not always the case that the responsible configurations are making all the decisions concerning their cases. It is not that unusual for a decision to be taken at the first up-coming Council meeting even if this is not the sectoral configuration to which it belongs. This is often the case with regard to decisions that have been agreed upon at the lower levels of the Council hierarchy and been labelled 'A' points. Hence, we may ask whether there is a difference between 'A' and 'B' point items with regard to openness.[64] As with the EP, the separation of discussion and voting processes is a disadvantage from a transparency perspective, but it is not detrimental to the overall compliance with this indicator.

However, the Council has informal meetings prior to and in between the formal meetings in order for ministers to discuss cases in a more low-key format. Transparency of these meetings is obviously low as there are no minutes or reports available from such encounters (Hayes-Renshaw and Wallace, 2006: 49). After enlargement, the frequency of these meetings has seemingly increased (Hagemann and De Clerck-Sachsse, 2007). Hence what has been 'won' with regard to the introduction of public deliberations being made available on the internet has arguably been lost in the sense that the 'real' discussions take place informally out of the public eye, even though the Council is at its most open when dealing with OLP dossiers.

Rules ensuring and inducing equality among the participants?

The various Council configurations usually meet every four to six weeks. In contrast to the working parties and Coreper, the ministerial meetings are subject to more formal rules. Attendance, convening and the chairing of meetings (i.e. the rotating presidency) are regulated in Articles 16 (TEU) and 237 (TFEU) and followed up in Article 1, 5 and Annex I in the Council's rules of procedure. Only representatives at the ministerial level can meet as full members, but the member states decide themselves which minister to send. In addition, and as mentioned above, the Commission and lower level officials may attend. It is mainly the Council President or the Commission that can initiate ministerial meetings. Upcoming meetings and agendas shall be announced in advance both on an six- and eighteen-month basis (see Articles 1 and 3). Moreover, the Presidency is responsible for ensuring that the agenda and accompanying documents are sent to all the members at least 14 days in advance (rules of procedure, Article 3). Hence, the member states must announce items for the agenda 16 days in advance. The rules of procedures also describe the division of the agenda into 'A' and 'B' parts signifying those items that do and do not require discussion respectively. The general framework for ministerial meetings is regulated in Article 5 and basically states that these meetings shall be conducted in openness during legislative procedure. In comparison, non-legislative sessions or 'In other cases, meetings of the Council shall not be public' unless the Council or Coreper so decide (Article 5, see also Article 8). As we have seen, more specific transparency rules for the OLP are covered in Article 7 of the rules of procedure. Article 7 also obliges the Council to provide an explanation of votes. Many of these formal rules will arguably help to level out the playing field between the member states.

In the meetings, the Presidency has a leading role as it convenes and chairs the meetings, including the informal bilateral and multilateral talks. This role, in addition to its participation in the inter-institutional trialogues with the EP and Commission, puts the Presidency in a powerful position as it has an information advantage in relation its fellow members which can be used to manoeuvre the discussions in the desired direction (Bjurulf and Elgström, 2004: 261). However, given that the Presidency is (as is no longer the case in field of foreign affairs) a rotating office, every member state gets their turn at the wheel.

Due to enlargement, the time for debate in the ministerial meetings has decreased, which may arguably make the relationship between the member states less equal as the discussions take place outside the formal meeting room. In their study, Hagemann and De Clerck-Sachsse (2007: 12, 14) found that after enlargement more dossiers were contested, but without voting against the majority. Instead, the usage of formal statements/explanations attached to the Council minutes has increased:

> One such measure is the use of formal statements, which are included in the official Council decision records, to signal a country's reservations or even outright opposition to an act, even in cases where this was not expressed through voting. Hence, while the Council records now show an even-greater degree of consensus as far as voting behaviour is concerned, the governments are at the same time able to ensure the recording of their political positions in these formal statements. The use of formal statements has risen considerably since enlargement, and the proportion of legislation adopted in 2006 with one or more governments voicing reservations either through direct voting or in the formal statements amounted to a surprising 45%. Adopting legislation in the Council hence seems to be a two-sided political game that is increasingly accepted even in the official records.
>
> (Hagemann and De Clerck-Sachsse, 2007: 34)

The Council ministerial meeting is an official decision-making setting and the rules for qualified majority voting are extensively documented both in the Treaties (Article 238, TFEU) and in the rules of procedure (Article 11 and Annex III). However, the consensus reflex is strong and even though the rules allow for qualified majority voting, as is the case under the OLP, decisions are usually not taken before everyone agrees (Heisenberg, 2005). This consensus culture is based on the principle of 'diffuse reciprocity', 'in which member states agree to respect each other's vital interests and not to create structural minorities that are consistently overruled in decisions' (Farrell and Héritier, 2004: 1195). Mattila (2004, 30–1) describes the process when a dossier reaches the ministers like this:

> A proposal can be adopted with a single sentence from the Chairperson when he or she knows that there is unanimity in the Council. If the Council is not unanimous, the Chairperson still knows the Member States' positions. If he or she decides that enough Member States are on board, the proposal is accepted and those opposing or abstaining from the decision can record their views officially.

Moreover, the Presidency chairperson may declare that agreement has been reached 'without being contradicted' (Hayes-Renshaw and Wallace, 2006: 269). This type of implicit voting (together with agreements made on the lower levels) makes it very hard, of course, for the untrained eye to detect where the differences between member states lie unless, as mentioned above, member states attach a statement to the vote.

With enlargement and a significant increase in the number of ministers around the table, the meetings have become more formalised and streamlined which means that many substantial discussions take place more informally (more on this below). Hence ministerial meetings are characterised by more informal sessions between the formal plenary meetings where the Presidency conducts bilateral and/or multilateral talks. Or according to an interviewee in the Council:

> Negotiations now take place elsewhere, presidencies have bilateral contacts with people … it happens in a much more informal manner outside the meeting room, in bilateral or multilateral contacts … so the presidency has to go round and hear all the major member states and actually push them to see how far they will go and that helps to put together a compromise proposal … the bilaterals have, in a way, become more important than the plenary sessions.
>
> (Hagemann and De Clerck-Sachsse, 2007: 24)

Moreover:

> enlargement has brought about … important atmospheric changes related to the work processes and conduct of negotiations in the Council. It was reported that our 52 interviewees (though, of course mainly the interviewees from the old members commented on this) have experienced an increasing impersonality in the style of the meetings. The legislators do no longer necessarily know each other, let alone are able to connect across an increasingly longer negotiation table. Some observers therefore emphasise the increasing tendency towards 'statement reading' in the Council formations rather than actual negotiations. This has been seen as one of the reasons why negotiations have now moved elsewhere.
>
> (Hagemann and De Clerck-Sachsse, 2007: 39)

These developments could have implications for internal solidarity and the willingness to abide by the aforementioned principle of 'diffuse reciprocity'.

Time for debate prior to decision-taking?

As we have seen, the Council ministerial meetings follow the structure of 'A'and 'B' points. Hence, the time allocated for debate on dossiers appearing as 'A'points is already reduced. Any 'B' points on the Council agenda warrant discussion, but it is often pointed out that the amount of 'B' points is much lower than the number of 'A' points as the majority of items are decided in the lower echelons of the Council hierarchy (working groups and Coreper) (see Hayes-Renshaw and Wallace, 2006: 58) and the share of 'B' points has reduced even further after the last enlargement. In his study, Settembri (2007: 29) found that whereas the number of dossiers adopted as 'A' points has increased slightly, the usage of 'B' points has decreased significantly and 'is particularly pronounced for major acts'.

Moreover, enlargement has also had an impact on how the Council now conducts its meetings (see Article 20 and Annex V in the rules of procedure). An increased number of member states implies more speakers and thus more time pressure during the formal meetings. A typical Council meeting starts with the Presidency presenting the issue, then usually giving the floor to the Commission representative who goes more into details about the legislative proposal. Earlier, this used to be followed by a 'tour de table' giving each minister time to state their opinion on the dossier. It is this practice that has more or less been abandoned or rather modified so that only those members opposing the proposal are asked to speak. For instance, the Council rules of procedure now states that the Presidency is encouraged to:

> (c) organise the time allotted for discussion of a particular item, in particular through limiting the time during which participants may speak and determining the order in which they may take the floor; (d) ask delegations to present in writing their proposals for amendment of a text under discussion before a given date, together with a brief explanation if appropriate; (e) ask delegations which have identical or similar positions on a particular item, text or part thereof to choose one of them to express their shared position at the meeting or in writing before the meeting.
>
> (cf. Article 20(1c–e), Annex V, Council rules of procedure, see Council, 2009)

This trend is not very conducive to deliberation, especially if we take into consideration that there is:

> A tendency towards reading out statements and keeping to the formal procedures of negotiations has prevailed.... The formalised tone and structure that now characterise the meetings and work culture have led to an

> emphasis on the formal statements instead of merely voicing either concerns or opposition on a more casual basis.
>
> (Hagemann and De Clerck-Sachsse, 2007: 12, 14)

Following some of these sessions, most representatives usually read prepared statements out loud and there are seemingly few/no spontaneous responses to the positions presented (see also Hayes-Renshaw and Wallace, 2006: 125–6). However, having said this, Bauer (2004: 373) argues that 'Nevertheless, the majority of public debate on "B" items which have taken place are not mere "shows" of pre-cooked agreements for the public, but real discussions'.

But there are also other obstacles to debate in the Council, as the participants come and go during the formal meetings and are also disturbed by

> Whispered conversations within and sometimes between delegations.... It demands real power of concentration and stamina for the participants to listen carefully, as well as to speak, not to mention to observe and interpret the 'body language' of other participants.
>
> (Hayes-Renshaw and Wallace, 2006: 51)

Consequently, even if free discussion does take place, Council ministerial meetings seem to be organised and conducted in a way that does not seem very conducive to deliberation.

So, do the Council ministerial meetings comply with the requirements for a democratic deliberative meeting place? In the Council it is the working groups and Coreper that prepare the dossiers. In this sense, the sequencing in the Council is similar to that of the EP where it is also the parliamentary committees that prepare the report. However, the major difference is that the division of labour in the Council is not compatible with what should be expected from democratic legislators, as the elected ministers are hardly present. It is the unelected officials in the working groups and Coreper who do the collective preparatory work. In the EP, by contrast, there are always MEPs at both the committee and the plenary levels (I come back to this under the next criterion). For this reason the Council's way of organising the preparation of its policy position is not compatible with a democratic opinion- and will-formation process. This is also underlined by the fact there are many implicit rules governing the ministerial meetings that are not accessible and self-explanatory for citizens and other interested parties (cf. Lewis, 1998, 2000).

The Conciliation Committee

Elected participants?

The Conciliation Committee is the first and only *formal* setting regulated in the Treaties where Council and EP representatives meet face to face under the OLP (Article 294, 10–12: TFEU).[65] The Committee consists of an equal number of participants from the EP and the Council (27 each), i.e. their respective conciliation delegations. In reality, however, a lot more than 54 people meet in the Conciliation Committee as the delegates also bring staff members to assist them (political advisers, technical and legal experts, other supporting staff members etc.). Whereas the EP, as we have seen, appoints a new delegation for each dossier, the Council delegation to the Conciliation Committee is always the same, that is, the permanent representatives and the minister from the member state holding the Presidency (hence one representative from each member state).[66] In addition, 'The relevant Commissioner usually represents the Commission' (Corbett *et al.*, 2007: 224). At this stage, the Commission does not have the right to vote on or withdraw a proposal,[67] but functions as a mediator and facilitator for reaching agreement between the EP and the Council.[68] In sum, whereas the EP's delegation consists of directly elected MEPs reflecting the political balance in the Parliament, the Council has, at best, only one representative with a popular foundation. This makes the Conciliation Committee a setting with a 50-50 share of participants representing affected and competent parties respectively. It is thus difficult to determine for which category of meeting places it qualifies and additional information is therefore needed.

Access to documents and debates?

All Conciliation Committee meetings are closed to the public when they are in motion and information is not accessible before the process is over. This rule also covers access to the strategic documents. Neither the working documents nor the other preparatory documents used during conciliation meetings are publicly accessible prior to and/or during the process.[69] Afterwards, however:

> The joint text agreed in conciliation is posted in a provisional version in one language (subject to legal-linguistic revision) on the Parliament's website as soon as possible after the end of negotiations, thus enabling the general public as well as the representatives of the institutions to assess the outcome.
>
> (EP, 2007b: 18)[70]

The Conciliation Committee does not have an internet site of its own, but other documents that are publicly available can be found at the EP's conciliation and co-decision website, which is the only place where information on conciliation issues are collected and stored systematically.[71] Here one can find information about the members of the Conciliation Committee, what conciliation and co-decision are, summaries of on-going dossiers broken down on an individual basis from procedural history, background papers to official texts as well as activity reports and a co-decision newsletter with monthly updates and summaries of conciliation proceedings. The Council hardly provides any information about the Conciliation Committee, let alone its proceedings. In sum, access to documents and debates are complied with on an ex post basis due to the EP's conciliation and co-decision website.

Rules ensuring and inducing equality among the participants?

Unlike other settings during the OLP, the Conciliation Committee is described in Article 294 (10–12). Here, both membership (cf. above) and the competencies of the Committee are defined. The latter involves the right to negotiate and to vote.[72] The Conciliation Committee is thus an official decision-making setting. In fact, the powers of the Conciliation Committee are further elaborated upon by the ECJ with regard to whether amendments beyond what has already been suggested in the second reading can be considered. The Court ruled that 'The wording of Article 251 EC [now Article 294 TFEU] does not ... include any restrictions as to the content of the measures chosen that enable agreement to be reached on a joint text' (cited from the EP, 2007a: 7).[73] Hence new amendments can also be tabled.

The Commission is also present, but has no voting rights. It rather functions as a broker aiming at facilitating agreement.[74] In this sense, there are some actors (the MEPs and Council participants) who are more important than others (the Commission). However, the Commission is not a legislative body and the Conciliation Committee is convened as a last resort to reach agreement between the EP and the Council as the two legislative houses. Hence it would not be legitimate for the Commission to have rights on a par with the EP and the Council. In addition, we may say that the EP and Council chairs possess key positions (see below).

The Joint Declaration fleshes out the treaty article in detail. To ensure equality between the EP and the Council, the two legislative bodies alternate with regard to chairing (EP Vice President and Presidency minister) and hosting the meetings (rules 28 and 32). The EP and Council chairs together also set up the agenda, the dates and the items for Conciliation Committee meetings (rules 29–30). In sum, we may say that the Conciliation Committee arrangement is compatible with the principle of equality.

Time for debate prior to decision-taking?

In Article 294 it is stated that the Conciliation Committee shall seek to reach 'agreement on a joint text ... on the basis of the positions of the European Parliament and the Council at second reading.' Hence, the purpose of the inter-institutional Conciliation Committee is to reach agreement on a Joint Text, which the Council and the EP then take back to their respective institutions for final approval or rejection in the third reading.

To facilitate cooperation between the EP and the Council, it was agreed that they should work from a common document in order not to waste time on talking past each other (Shackleton and Raunio, 2003: 177). The joint working draft has an 'A' and 'B' part: 'part A contains the elements of the compromise already agreed during the preparatory work [i.e. through trialogues] and part B the unresolved points with the respective negotiating positions' (Cortes, 2000).[75] Hence, the Conciliation Committee – as the Council – works according to the 'A' and 'B' point procedure where time for debate is reduced to the latter points and as many points as possible are resolved in trialogues prior to or in between the meetings of the Conciliation Committee (see EP, 2007b: 16–8).

However, as the name of the committee indicates, its purpose is to reconcile the remaining differences between the EP and the Council after two unsuccessful readings, where there is still hope of reaching an agreement. This situation arguably focuses the Conciliation Committee meetings on negotiations rather than discussion and debate. The positions are elaborated and well known through discussions and amendments in the EP report and the Council's Common Position. However, the perception among the participants is that it is too difficult to reach agreement in a room with over 100 people. Hence, the Conciliation Committee has proved to be rather ill-suited for the above purpose. In short, this has made the Conciliation Committee more of a final voting arena than a conciliation and bargaining arena, as the more substantial inter-institutional discussions are instead taken in the trialogues. This does not mean that the Conciliation Committee is reduced merely to voting, but as we have seen above, 'The discussion is normally limited to these outstanding [B points] issues' (EP, 2007b: 17; Rasmussen, 2005: 1025).

From the above it becomes clear that the Conciliation Committee is neither designed nor intended to be, nor does it function as, a setting where OLP dossiers can be tested and justified in a publicly accessible manner. Instead, we could argue that the Conciliation Committee is designed as a bargaining and voting arena in the sense that it is convened at the end of the process when positions are known and consensus has not yet been reached, but where there is still hope for an agreement if both parties are willing to give and take. This does not make it democratically illegitimate, as in principle it seems to serve a perfectly legitimate cause in providing a shielded setting where the parties can discuss and potentially agree on outstanding points of disagreement. It is legitimate because it appears towards the very end of the procedure when the positions are known, but where the process needs new momentum in order to potentially reach settlement on a Joint Text. However, due to the growing importance of trialogues, discussions in the Conciliation Committee have become less important. In addition, and despite not being illegitimate as such, it does not qualify for the category of a democratic deliberative meeting place.

Conclusion

The first criterion – deliberative meeting places – seeks to identify those settings where it can be possible for OLP dossiers to meet critical argumentative scrutiny in a manner that is openly accessible to citizens and affected parties. The existence of democratic deliberative meeting places is thus a necessary condition in order for representative systems to rely on the legitimacy of procedure instead of actual citizen participation without breaking the promise of government by the people. Failing to provide such meeting places (without replacing them with similar mechanisms) would hence render the polity to merely governance for the people.

However, this does not mean that the OLP process needs to consist of such meeting places alone. Rather, the first criterion requires that each legislative body should provide at least one setting where it can be assumed that arguments can be raised, discussed and justified in a publicly accessible manner. A minimum requirement for such a setting is that the participants are popularly elected and, further, that it is composed in a manner that can be deemed representative of the institutional actor they represent. In this regard, five OLP settings were considered: the EP committees, the EP plenary sessions, the EP conciliation delegation meetings, the Council ministerial meetings and the Conciliation Committee meetings – although the latter with initial reservations due to the fact that only half of the participants are popularly elected.

The evaluation showed that the EP committees and plenary sessions are both settings where it is likely that there can be an open testing of different positions on OLP dossiers. Both settings are composed of elected representatives – the committees of a representative selection of the plenary. Meetings are conducted in openness, and there is access to both debates and documents, although the transparency requirement could have been managed more efficiently in committee with regard to verbatim records as is offered during plenary sessions. The recent development of video streaming of committee meetings is definitively an improvement in this regard. From an openness perspective, however, both settings lack satisfactory voting procedures as roll call votes (which are an optional choice) are only conducted upon demand. Instead, most votes are taken without a full record of names. There is also an extensive set of rules to ensure equality among the MEPs, especially for the plenary sittings, but also for committee meetings. These rules range from speaking time to the right to change the agenda, the right to suggest amendments to the report etc. It should, however, be noted that the formal rules as well as the informal practices favour the three largest political groups, to the disadvantage of the small political groups and especially the non-attached MEPs.

Due to the number of participants, speaking rules and time limitations, it was argued that the types of deliberation that could be expected in the EP committees and plenary sessions respectively were probably different. In other words, the committees are smaller and less formal than the plenary sessions and thus have more time for discussion. Moreover, the committee is the setting where the draft work is done whereas the plenary sitting considers the report prepared by the committee. Presumably, these aspects make the committee a setting more conducive to the argumentative exchange of positions than plenary sittings. In this connection, I argued in line with Goodin (2005) who suggests that under non-ideal conditions we should think of, and expect, a sequencing of the deliberative tasks.

The EP conciliation delegation meetings were also appraised. In many respects they comply with the requirements for a democratic deliberative meeting place since they are composed of a more or less representative selection of elected representatives (although non-attached MEPs are no longer part of the delegations), meeting documents are available (although only ex post) and the meeting rules are designed to ensure a certain level of equality among the MEPs. In addition, there is time for debate. When the EP conciliation delegation meetings do not fully meet the expectations of a democratic deliberative meeting place, it is because they are not intended to be open settings for testing and scrutinising arguments and positions. Rather, meetings are deliberately held in camera in order not to give away bargaining points to the Council prior to the conciliation trialogues. Moreover, it is a meeting place where the MEPs discuss strategies and provide the trialogue-MEPs with tasks and mandates for the final negotiations with the Council. This does not make the setting illegitimate, it only means that it cannot be considered as a contribution towards ensuring the existence of democratic deliberative meeting places for the production of OLP acts.

The evaluation of the conditions on which the second legislative house, the Council of Ministers, is organised revealed that this setting also complies with many of the requirements of a democratic deliberative meeting place. It is composed of a representative selection of (even if only indirectly) elected representatives and the meetings are now conducted in more openness than they used to earlier. Though video streaming and minutes are available, verbatim records are not. In addition, provisional agendas, voting charts, monthly summaries and press releases are accessible from the Council website. However, the problem is that the ministerial meetings are not the setting where the preparatory Council work is conducted. This is done in the working groups and in Coreper, and neither of these settings are open to the EP or the general public. Moreover, Settembri (2007) also found that since enlargement, there have been an increasing number of informal (bilateral/multilateral) meetings in between the formal ministerial meetings. He also found that enlargement has seemingly significantly reduced the number of 'B' points discussed at the ministerial level and the number of 'A' points has increased slightly. In sum, it is highly questionable whether these features make it likely for ministerial meetings to provide the necessary conditions for open and critical scrutiny of OLP acts.

The Conciliation Committee is neither intended, nor does it function, as a deliberative democratic meeting place. Rather, and as the name of the committee indicates, it is a meeting for settling outstanding issues towards the end of the legislative process in order to exhaust all possibilities for reaching an agreement. Formally, the intention behind the Conciliation Committee is presumably to be a bargaining or a final conciliation setting, but the impact of the trialogues has skewed its role more towards a final clearing or voting arena where what has been agreed in trialogues is formally approved in the Conciliation Committee.

In sum, then, it is only the EP committees and plenary sessions that comply with the requirements for a democratic deliberative meeting place. Since the Council ministerial meetings fail in this regard, it cannot be argued that the OLP fully complies with the first criterion.

6 Inclusion of affected and competent parties

Introduction

There is no foundation for democratic government without the inclusion of the views of affected parties. Representative democracy always faces the problem of finding a mechanism to convert the slogan 'by the people' into an acceptable reality in the eyes of the citizens, as the latter are never actually present when decisions are taken. In this chapter, we will look more closely at the type of *actors* who are most involved in co-decision-making. More precisely, we will take stock of how the balance between affected and competent parties patterns. The first indicator refers to the *institutional actors* and requires that the main decision-making bodies are popularly elected and that they – as collectives – possess veto power. The second indicator is oriented at the level of *individual actors* and seeks to determine whether co-decision-making respects the normative hierarchy between representatives of affected and competent parties inside the Council and the EP. In addition, the Commission's consultation system is also assessed in order to find out whether the inclusion of civil society actors in the preparatory process can somehow compensate for the Commission's lack of popular anchoring.

Elected decision-makers with veto power

> Indicator: the main institutional decision-making bodies are *popularly elected* and have collective *veto* power over the final outcome.

The first indicator of inclusion demands that the main, and most important, decision-makers are popularly elected representatives situated in bodies that have veto power. According to Article 294, the main institutional actors in co-decision-making are the Commission, the EP and the Council, but it specifically stipulates that it is the EP and the Council that jointly exercise the right to either accept or reject a proposal, that is, both institutions have to agree in order for an act to pass. Hence even if the Council is only indirectly elected, and many hold that the EP elections are only second-order (Reif and Schmitt, 1980), the power to veto a proposal is in the hands of two bodies with a popular mandate.

As we have seen, the Commission does not have the final say on OLP dossiers – it cannot formally veto decisions on a par with the EP and the Council. This point notwithstanding, the Commission is a powerful actor and wields considerable influence especially due to its monopoly to formulate and table proposals. What is more, during the first and second readings, before the conciliation process starts, the Commission must also be consulted and give its opinion on the amendments made by the EP. The Council can only adopt the EP amendments the Commission rejects if it is able to drum up unanimity. In some way we may argue that the Commission also possesses a form of veto with regard to its right to withdraw legislation before the conciliation process starts, that is, 'If it considers the final consensus at Council level as conflicting with its proposal, it can withdraw this and thus exert veto power' (Van Schendelen, 2006: 34–5).[1] In addition, the Commission acts as a facilitator and broker throughout the process and is the only institutional actor that has access to all the central settings both in the Council and the EP.[2] Even if it is short on veto power, the Commission has strong agenda-setting powers, especially in the preparatory phase prior to the official tabling of a dossier and to some extent during the formal procedure through its gate-keeping role.

The Commission's lack of an electoral link is consequently a democratic shortcoming, especially when there is no real balance of power dynamic between the Council/EP and the Commission. Having said that, it should be remembered that however important the Commission is in setting its stamp on OLP dossiers, it is nevertheless only the EP and the Council that have final veto power to turn a dossier down and both these institutions have (at least) formal direct and indirect popular anchoring respectively. In this respect, then, we may say that the first indicator requiring that the main decision-makers possessing veto power must be popularly elected, is complied with. But from a deliberative perspective, this in itself is not enough to determine whether the OLP fulfils the procedural qualities of democratic inclusion. In emphasising the process leading up to final decision-making, we also need to know which individual actors shape the will-formation process and how the balance between representatives of affected and competent parties patterns more precisely.

Hierarchy of actors

> Indicator: the *elected representatives* are the *deciding actors* when *discussing and deciding* how consequences and burdens are distributed. It is required that elected representatives sit together and *collectively* mould an opinion and discuss a case prior to decision-taking. The involvement of *competent parties* is limited to preparing policy-making.

The second indicator of inclusion refers to the overall involvement of the various individual actors operating within the decision-making bodies and seeks to establish whether the normative hierarchy between affected and competent parties is respected. In other words, is there 'vertical accountability' between elected and non-elected representatives under the OLP?

The Council

From the previous chapter on meeting places we already have an impression of the relative importance of the various OLP actors. In the Council, the lower level bodies (working parties, Coreper, Secretariat) play a significant role for the preparation and articulation of the Council position. It is therefore pertinent to address how the internal relationship between the working parties and Coreper, on the one hand, and the ministerial level, on the other, affects the Council's status as an institutional actor with (indirect) popular anchoring. The question is whether the indirect electoral link is still 'valid' given the heavy involvement of the non-elected officials in the legislative process. In other words, to what extent can it be argued that the national ministers can also be seen as the citizens' representatives at the European level when they are acting in the Council and thus validate the intergovernmental view that the Council is a democratising factor in the EU?

My conclusion, as we shall see, is that they cannot. The reason for this is, as expected, due to the indirect nature of democratic representation in the Council, but not because deliberative democracy is necessarily incompatible with indirect representation as such. The problem with the indirect chain of representation in the Council is that this is not the primary work setting of national ministers, and thus their jobs as ministers in the member states are too time-consuming in themselves to be compatible with the time and attention needed for legislative work at the European level (cf. Chapter 5). By this I do not mean that the Council as such neglects its duties and tasks. The problem is the extent to which the ministers are forced to neglect their duties as *democratic* representatives in the Council since they must delegate the majority of the work on legislative dossiers to Coreper, the Secretariat and the working parties.

More precisely, the Council's composition of national ministers may have a sound intergovernmental or national logic, but it does not withstand the logic of democratic representation in a deliberative sense as long as the national ministers do not really have the time to sit together and collectively mould and discuss legislative proposals prior to actual decision-making. The difference between the EP and the Council is here striking and illustrative. MEPs are 'hands on' throughout the legislative process and do most of the work themselves. And, more importantly, they sit *together* first and foremost in committee, but also in plenary sessions to discuss and find solutions.

In the Council it is the working parties, Coreper and the Secretariat that take care of the day-to-day business concerning OLP dossiers, including the contacts with other actors. As mentioned above, OLP dossiers have been prepared first by the working parties and then filtered and further refined by Coreper before they reach the ministers. 'Insiders' to the Council system estimate that 85 per cent of all decisions taken in the Council are not subject to any discussion, but simply adopted without any further ado as 'A' points at the ministerial level.[3] Now, it may be objected that the work of the Council officials is hardly a 'one-man-show' where they can operate independently of or sideline the ministers – at least not without repercussions. Rather, they are highly dependent on retaining good working relations and trust with the principals in their respective capitals (Hayes-Renshaw and Wallace, 2006: 80). Although the 'A' and 'B' point procedure says something about the importance of the preparatory bodies, Hayes-Renshaw and Wallace (2006: 79–80) underline that a dossier – and in particular a controversial one – is usually sent up and down the Council hierarchy before a final decision/vote is taken. In other words, when a final decision is to be taken at the ministerial level it is not the first time the ministers hear of the case. In addition, Hayes-Renshaw and Wallace (2006: 48) also point to the fact that ministers have to prepare themselves prior to the meetings and thus that 'the heavy and repeated engagement of many national ministers in the Council reveals a pattern of transnational involvement that marks the EU out from other forums for negotiations between governments'.

However, the permanent representatives are *not* at all times following strict orders from member states. Given the permanent character of Coreper, the shifting Council configurations as well as the rotating nature of the Presidency, Coreper, in particular, represents a much needed continuity, expertise and institutional knowledge. Often – and especially with regard to smaller member states – there is initially no official opinion and the permanent representatives are more or less free to form a member state's position (e.g. Lewis, 1998: 490–1). This obviously creates the potential for a considerable amount of manoeuvring and influence (Rhinard, 2002: 202).

Moreover, the problem is that issues labelled as 'A' points may be known to the ministers individually, but they have hardly been dealt with by them as a collective body before a final decision is taken. The whole point of the 'A' point procedure is to relieve the ministers of their work pressure. The legislative process always starts below the ministerial level and if agreement can be found here, the issue is moved up to the ministers where it is formally decided upon at the first upcoming Council meeting, regardless of whether this is the Council configuration in charge of the given dossier. Here, the minister 'rubberstamps all these A-dossiers in one minute' (Van Schendelen, 2006: 34). In this sense we could argue that at the minister level decision-making is most of the time merely the result of an aggregation of preferences. In contrast, it is at the working group and Coreper levels that decision-making is characterised by a collective process resulting in more deliberatively reached agreements.

What is more, the working groups and Coreper are arguably not just preparatory, but also decision-making settings (Rihnard, 2002: 195). This is also more or less confirmed by Fouilleux *et al.* (2005: 614). In interviews with Coreper, the Secretariat and working party officials, they found that only the respondents mentioned concretely that budget issues as well as 'deal-making involving intersectoral trade-offs is left for ministers to handle' (ibid). Moreover, Settembri (2007: 29) also found that after enlargement, the number of acts adopted as 'A' points has increased slightly, whereas the number of acts subjected to 'B' point discussions has decreased significantly, especially for important dossiers. In other words, enlargement seems to significantly affect the time allocated for debate. Settembri (ibid) also looked at the balance between ministerial and diplomatic representation when decisions were taken. Since enlargement this balance has tilted even further away from the ministers.[4]

It may be that this procedural set-up is a 'sensible solution to the problem of ministerial overload' as Hayes-Renshaw and Wallace (2006: 80) argue, but from a democratic perspective it illustrates clearly the problem of part-time ministers. True, the ministers can at any time convert 'A' points into 'B' points and thus demand discussion on the topic. They can also veto the proposal, as they are the only ones with formal decision-making power. This point notwithstanding, it is the lower level officials and not the ministers who have been part of a thorough *collective* moulding process prior to final decision taking. From a deliberative democratic point of view, it can therefore be argued that the meetings among ministers in the various Council configurations are more often than not reduced to rubber-stamp settings. Hence it can also not be argued that the ministers function as representatives who voice the views of affected parties. The 'A' point procedure may improve efficiency, but it does not improve the *democratic* quality of decision-making and, consequently, it does not respect the normative hierarchy between affected and competent parties. In the Council it is the unelected officials (the working groups, Secretariat and Coreper) who are the real movers of OLP dossiers and the ministerial level is more or less reduced to a setting that adopts what has already been decided at the lower levels. In this way, the organisational set-up or division of labour inside the Council upsets the democratic connection between opinion- and will-formation processes, on the one hand, and final decision-making or voting procedures, on the other.

The EP

With regard to the EP the situation is different. Unlike the ministers in the Council, the MEPs do the preparatory work themselves – the EP Secretariat helps out, but does not do the political work for them.[5] In this sense, the EP's organisation of legislative work respects the hierarchy between affected and competent parties' representatives and ensures incentives for affected parties' representatives to sit together and *collectively* shape a position.

Moreover, if we look at the EP committees and the plenary sessions together, the division of labour between the two types of settings is also compatible with a democratic deliberative perspective. The EP – first in committee(s), then in plenary settings – conducts the preparation and discussion of OLP dossiers in a publicly accessible manner. In the previous chapter, it was argued that the way the EP is organised is conducive to democratic deliberation and hence also to the inclusion of affected parties' views. So even if the major work is done in the committees, this is (in principle) not a problem as the composition of EP committees follows the numerical strength of the various political groups and can thus be deemed representative of the EP in plenary (McElroy, 2006: 25). When seen in isolation, then, the division of labour between committees and plenary sessions makes sense not only from a practical perspective, but can also be defended from a normative point of view.

However, the EP does not operate in a vacuum and when we change the perspective from how the EP is organised internally to how the inter-institutional co-decision-making system seems to affect the EP's working methods, a more dubious pattern appears. The committee has always been an important setting in the EP. However, the trialogues seemingly help to undermine the importance of committee work and consequently also downgrade the committee as a central legislative setting. Earlier, all the substantial discussions took place in the formal arenas such as the committees and plenary. Now, many of these discussions take place in trialogues. Although the committee is obviously not void of importance, the trialogues seriously challenge it as being *the* place where the EP makes its major policy choices. Moreover, when the majority of dossiers are now the result of 'early agreements', this enhances the importance and impact of first reading trialogues to the detriment of the committee, which is the only setting open to public scrutiny in the very beginning of the formal decision-making process. Let us have a closer look at how the trialogues affect the internal organisation of the EP.

The trialogues are exclusive to the participants. Hence other MEPs are dependent on the information provided by the trialogue-MEPs. Those who do not have access to trialogues only learn what has been agreed afterwards and are seemingly then often left with a choice of either approving or rejecting the agreement. And the rejection does not come without costs, as the MEPs voting down the proposal are then responsible for the failure of the legislation. According to Farrell and Héritier (2003b: 24) when agreement in first reading 'works successfully, the Parliament and the Council do little more than sign off on a deal that has already been negotiated among a small group of actors'.

This situation is not completely uncontroversial in the Parliament: 'The basic problem is that the positions expressed to the Council during such "unlawful" contacts do not necessarily represent the points of view of the majority of the members of the committee concerned.'[6] Fouilleux *et al.* (2005: footnote 23) reported that in an interview with a Commission official it was argued that:

> The problem is that this procedure is a little too quick and not transparent enough. When we reach agreement on the first reading, the essential work is done between the President of COREPER and the EP's rapporteur. It is up to the rapporteur to consult the other members of Parliament ... So MEPs are often confronted with the choice between 'yes' and 'no' – there is no deliberation.

Moreover, the first reading is different from the others in the sense that the EP and the Council have not yet formed a formal position and this gives the key actors considerable room for manoeuvring as there are no limitations on how many and what kind of amendments can be presented at this reading (in contrast to the second and third readings) (Rasmussen and Shackleton, 2005: 12).

The problem with first reading trialogues is that until the Council has formally adopted its Common Position there are no time limits, so the process of moulding and shaping a policy proposal can go on for a long time in secret trialogue meetings before they enter the formal decision-making arenas (Farrell and Héritier, 2003b: 23). Moreover, 'Negotiations on early agreement dossiers are almost entirely informal – it is often extremely difficult for others within the Parliament, let alone outsiders, to have any idea of what exactly is going on in a specific brief' (Farrell and Héritier, 2003b: 8). Knowing how important the first stages of a decision-making process are, this is, of course, problematic. But there are also critical comments to be made concerning *second* reading trialogues. Even if the OLP process is subject to strict time limits after the adoption of the Council's first reading Common Position, and even if the positions of the EP and the Council are known by then and the possibility of proposing amendments is constrained in scope, the problem is nevertheless that the:

> Council interrupts Parliament's second reading with a view to persuading it to adopt amendments that it (Council) can accept, thereby making conciliation unnecessary. In this case, it has become common for Council and Parliament to meet after the committee responsible has voted and to seek to negotiate compromises. At the very least, such an approach poses a challenge to the role of the parliamentary committee meeting as the place where interests are aggregated. Decisions are effectively being taken elsewhere in a body meeting far from public view.
>
> (Shackleton and Raunio, 2003: 178)

As mentioned, first (and second) reading trialogues are not undisputed within Parliament (cf. EP, 2007b, 2009b). Hence in 2004, general guidelines for how the negotiations with the Council should be conducted were outlined by the EP. These guidelines were later 'upgraded' to the 'Code of Conduct for Negotiating Codecision Files' in 2008.[7] In short, the Code encourages the participants to, among other things, announce trialogue meetings in advance, to appoint a balanced trialogue group, to seek a mandate for negotiations in the committee before entering into trialogues and to consider the need for early agreement on a 'case-by-case' basis (EP, 2009b). A so-called 'cooling off period' (of one month) between committee and plenary stages has also been encouraged by the Conference of Presidents, but the rule has 'not been consistently applied'. (EP, 2009b: 25). Moreover, neither the guidelines nor the Code is legally binding, but as the EP (2009b: 26) states itself: 'In comparison to the rather vague term "guidelines", its name is intended to underline its more binding nature.' (EP, 2009b: 26). However, even if 'early agreements are no longer struck in the kind of institution-free environment that existed earlier' (Toshkov and Rasmussen, 2012: 9), in the end the responsible EP committee only has weak means to hold the trialogue participants accountable during the first (and second) reading, especially if the Council is pushing for agreement.

The Code is inspired by, or based on, the 'accountability system' during the conciliation process where the trialogue-MEPs are controlled by the EP delegation. According to the EP itself, delegations keep close contact and monitor the trialogue participants. The latter receive their;

> mandate from the delegation before every trilogue, and ... report back to the delegation after each trilogue. Thus, it has been the delegation which has taken all important procedural decisions (e.g. whether another trilogue meeting should be arranged, or whether the Conciliation Committee should be convened). And it has remained the delegation which had determined the strategy vis-a-vis the Council position at every stage of the procedure; which has approved or rejected compromise proposals; and which, at the end of the conciliation procedure, and always by a vote in which the support of an absolute majority of members of the delegation (i.e. at least 13 votes) has been required, has formally approved or rejected the agreement reached.
>
> (EP, 2007a: 6)

Hence, they work on a mandate from the conciliation delegation and cannot take formal decisions, but have to refer back for approval the agreements reached in trialogues. However, it can be questioned whether the EP at large can sufficiently control the delegation. In fact, Rasmussen (2005: 1028) holds that 'it is considerably easier for delegations to monitor and sanction the key negotiators than for the legislative bodies to do so in regard to the delegations as a whole.' The problem, though, is that the process in the committee is in this way, and to a considerable extent, sidestepped and thus challenges the committee as the locus where collective opinion- and will-formation processes are conducted.

This emptying of the committee meetings has resulted in a fear (widely shared by the MEPs) that first reading trialogues in particular will reduce committee meetings to rubber stamp settings (Héritier, 2003: 827). As we have seen, there have been attempts at remedying this, among other things, by suggesting that:

> the committee should define a mandate to negotiate with the Council and then come back to the committee and report or that the Council should come to the relevant committee to present its ... common position and to negotiate with the EP before its second reading and not just talk to the rapporteur.
>
> (Héritier, 2003: 827)

As we know, the Council has not been willing to do so and prefers to attend the EP committee and plenary meetings as a listener. According to an MEP interviewed by Héritier (2003: 827), the Council has responded that:

> They do not want to come, partly because it is a public meeting, whereas a meeting with the rapporteur is a private meeting; nobody will report it later.

> Partly it is out of solidarity with the Council. They are not willing to say 'We are confident to get a qualified majority provided that X, Y, Z. It would give the EP, and indeed everybody, an opportunity to look into the Council'. Instead of such a committee, 'the Council would prefer to have more trilogue meetings with members of the political groups, a little representative sample of the committee put in a little room, which of course would not be open.

However, trialogue meetings are, as we have seen, not representative in the democratic sense. Moreover, and in order to anticipate where the majority in the EP will lie, the Council prefers to consult the large political groups to sound out and influence the result in Parliament. It is also the large political groups that usually sit on the key positions in committee such as the rapporteur, committee chair and shadow rapporteur, etc. To illustrate, the allocation of reports in previous parliamentary periods, Mamadouh and Raunio (2003: 346) found that 'the larger groups were over-represented (especially the EPP) while the smaller ones were under-represented.... only PES [now S&D] and EPP can normally afford the most expensive reports, such as those falling under the co-decision procedure' (see also Kaeding, 2005: 100). The political groups' control of the distribution of key positions means that non-attached MEPs in particular, but also those from small political groups, come out disadvantageously. These MEPs do obviously not have the same apparatus to take on the time-consuming work of being a rapporteur and will thus not be consulted by the Council or participate in trialogues. What is more, the requirement of an absolute majority supporting the EP's second reading position also results in consensus-seeking among the two largest political groups (EPP and S&D):

> Representatives of the two Groups meet with each other to strike deals over political or patronage issues without smaller Groups to left or right always being consulted. These latter may then be forced to conform on a take-it or leave-it basis.
>
> (Corbett *et al.*, 2007: 109–10)

These 'grand coalitions' are quite characteristic of EP outcomes (Lord, 2004: 119–20), but the process towards reaching them has been conducted without the non-attached and small political group MEPs taking part in it.[8]

Overall, small political groups increasingly find themselves excluded from the 'real debates' and negotiations. In an interview, a Green MEP admitted that 'as a small group, we would not be as involved as the larger groups (in formal trialogues) ... we are not part of that informal consensus' (Farrell and Héritier, 2004: 15). In other words, the moulding and policy-shaping deliberations to a large extent take place with less committee members being part of the process. Based on the above, it does not seem unreasonable to argue that if not all, then at least a significant amount of the real discussions now takes place informally and outside the committee framework. In this sense the smaller political groups have lost an important mechanism to exert influence over the legislative process, namely by formally proposing suggestions for amendments in committee (Farrell and Héritier, 2004: 16). Overall, the trialogues, especially if agreement is reached during the first, but also at the early second reading, 'threaten to short-circuit debate in committee and to create an élite amongst committee members who are privileged to deal with the Council' (Shackleton and Raunio, 2003: 177–8). This could, in effect, cause larger member states to 'to use their clout in Parliament to manipulate the legislative process in a non-accountable, and undemocratic fashion' (Farrell and Héritier, 2003b: 20).

To sum up, when the committee loses its central position as the core legislative setting where dossiers are discussed and shaped among a representative selection of MEPs, this affects the whole division of labour between the committees and the plenary sessions within the EP. Whereas the committee, like the trialogue, contains a reduced number of participants, the former has a democratic character, as its composition is representative of the plenary. As we have seen, to legitimise that the major chunk of legislative work is conducted in the committees is dependent on two factors: first, a representative composition of committees and second, that they conduct their work in a publicly accessible manner. On this basis it was argued that the plenary sessions can afford to be less probing and that the division of labour between the committees and the plenary sessions can be defended from a democratic point of view. Now that the most important discussions are conducted in the trialogues, this obviously also alters the assumed democratic quality of the EP settings (Farrell and Héritier, 2003b: 30). The paradoxical situation is that whereas studies have shown that the EP as an institution seems to be more influential in the trialogues than the Council (due to the latter's shortage of personnel and resources to follow *all* OLP dossiers all the way through the conciliation committee) this does not empower those MEPs not participating in trialogues: 'Thus while increasing the output legitimacy through more efficient decision-making, informal trialogues are counterproductive in terms of input legitimacy' (Häge and Keading, 2007: 358; see also Warleigh, 2003: 32). Chalmers *et al.* (2006: 155) put it even more bluntly: 'The trilogue is the biggest challenge to democratic legitimacy, for it centralises power in those actors who represent the Council and the Parliament at the trilogue' (see also Farrell and Héritier, 2003b: 8).

Even if the trialogue-MEPs are popularly elected and can thus be held to account in the next election, the possibility of holding them to account for what they have decided in trialogues suffers. In other words, their duty to explain and justify the positions and actions (defended in trialogues) towards citizens is not sufficiently respected, as trialogue agreements are often just presented as a 'take-it-or-leave-it' offer in committee. When trialogue-MEPs neglect this duty, they also cease being spokespersons voicing the views of affected parties. Hence the gain won in power is seemingly lost in democratic legitimacy due to the lack of representativeness of trialogues. In this way, it cannot be argued that their participation in trialogues contributes towards the respect of the hierarchy of actors because the collective opinion- and will-formation processes among elected actors is seriously challenged and is seemingly taken over by a small group of key trialogue participants – many of whom do not have any popular backing. This is even more the case when dossiers are completed as first or early second reading agreements, which is when the majority of OLP dossiers are finalised.

The Commission's consultation system

The Commission is also an important contributor to co-decision-making and even if its main problem (obviously) is that it has no popular anchoring, it wields quite an elaborate consultation system prior to the official launch of a dossier. This aspect should not be overlooked with regard to the question of inclusion and the relationship between representatives of affected and competent parties. The question is whether the inclusion of the various societal voices during the preparatory phase can compensate for the Commission's lack of a popular basis?

In fact, a tension between efficiency and legitimacy can be discerned in the Commission's consultation system. On the one hand, the Commission is a small organisation with scarce resources and personnel in comparison to the many tasks it has to fulfil.[9] To ensure efficiency and, at the same time quality, the Commission is therefore dependent upon external expertise ranging from scientific knowledge to information about the preferences of national governments. The expert committees play a pivotal role in the fulfilment of this end. On the other hand, the Commission is also in need of legitimacy which it has sought to ensure via complex processes of consultation with so-called stakeholders through mechanisms such as online consultations and policy and dialogue forums, as well as by including some NGOs and civil society organisations in the expert groups.[10]

Partly on this basis, it is commonly argued that the Commission is a rather open institution that facilitates a pluralist interest representation system and that is in contact with many types of groups (Mazey and Richardson, 2006a and b). There is no doubt that the Commission consults and includes a plethora of different interests, including many civil society and non-governmental organisations. We should, however, be careful about over-emphasising the democratic significance of consultation processes. Basically, the problem is a lack of representativeness and accountability. First, the individual delegates participating in the consultation processes are usually self-selected, that is, they have not been popularly elected, but are appointed by their organisation/group (at best, they are elected among their own membership). Self-selection is often biased towards those who are best organised, most educated, have more money and/or scream the loudest and to the detriment of those who do not possess these means/skills (Warren, 2008: 56). Moreover, interest groups approaching the EU level often lack a proper grassroots anchoring (Greenwood, 2003: 12).[11] It is also questionable how democratic it is to put too much emphasis on interest representation given that many civil society organisations only promote single issues. What is more, the participants themselves cannot be held to account for their positions by those who are going to be bound by them. Hence, 'At the system level, expanding the opportunities for participation can actually exacerbate deficits in democratic representation' (Warren, 2008: 56, see also Olsen, 2007: 130–1).[12]

Second, there is also discrepancy between the *principle* of democratic representativeness and actual *practice* within the group of *those who do have access* to the Commission's consultation system, as civil society involvement is only minor to other types of interests such as national government experts, scientific advice and/or business interests. The Commission expert groups play a far more important role than other consultation groups and the former consist almost exclusively of competent parties, that is, national experts and officials, scientists as well as private interests.[13]

The Commission is seemingly concerned with both the epistemic and the democratic aspects of consultation: 'good consultation serves a dual purpose by helping to improve the quality of the policy outcome and at the same time enhancing the involvement of interested parties and the public at large' (Commission, 2002: 5). It is nevertheless the Commission that orchestrates the consultation system according to its most pressing needs. The result is that the balance between those representing the views of competent versus those representing affected parties' views tilts towards the former. Hence the playing field does not ensure equal access. The consultation system has a top-down approach, where it is the Commission that draws and steers the pattern of interest group representation.[14] According to Menéndez (2009: 303), the problem is that:

> in the absence of functional European general publics, it is obvious that the Commission's strategy runs the risk of being the embellishment of lobbying, or truly a smart way to co-opt certain societal interests with a view of giving a legitimacy aura to technocratic decision-making.

Moreover, when the boundaries between the Commission and the interest groups participating in organised consultation processes are blurred, this also violates their democratising potential as:

> Participation is not about institutionalising protest. It is about more effective policy shaping based on early consultation and past experience.... As it is *participation by invitation*, the principle of representation is violated, and the Commission emerges as the sole representative of a European 'common interest'. Hence, governance comes down to *participatory engineering* as participation is mixed up with policy-making and implementation.
>
> (Eriksen, 2009: 163)

Technical versus political

There is also another obstacle reducing affected parties' opportunity to easily follow and access the political debate. This obstacle pertains to the practice of dressing up potentially controversial political questions in a technical language in order to speed up the co-decision-making process. The problem is obviously that a technical language often conceals the political salience and the normative dilemmas that a policy proposal may contain. Democratic politics require that such questions are dealt with in a language that spells out the normative dilemmas and that also clarifies value choices and potential solutions in an intelligible way for lay people such as citizens. This is the task of parliaments in particular, but also of other elected bodies such as the Council when acting in its legislative capacity. Legislative bodies deciding on proposals that will have binding effect on others have a responsibility to voice what is *politically* at stake in a policy proposal.

The lack of ministerial involvement in the Council has consequences for the democratic division of labour between elected and appointed actors in public policy-making because the latter also make political choices and not only technical ones. In their study, Fouilleux *et al.* (2005: 612) found that 'when one delves deeper into the way the Council functions, the distinction between technical and political issues is rarely so clear-cut.' There was also a tendency to define dossiers as 'technical' rather than 'political' in order to make them run through the Council hierarchy more efficiently. In other words, non-elected officials make important judgements that are not only of a pragmatic, but presumably also an ethical-political and moral character. Hence, the difficulty of separating technical and political questions and the constant blur of 'the frontiers between ... [working parties], COREPER and ministerial meetings' supports the above conclusion where it was questioned whether it is the ministers who are actually making the political choices. Moreover, given that politicians are generalists (even if many acquire special competence in certain policy fields) they presumably rely more on their advisers in the lower echelons of the Council than would have been the case had the Council membership been their full-time job (see also Fouilleux *et al.*, 2005: 610–1).

In addition to the highly technical nature of many dossiers, the domination of non-elected officials from the Council and the Commission also seems to affect the way debates are conducted among MEPs. In order to 'win credibility and legitimacy' with the experts especially in the Council working parties, the EP adopts 'the discourse of the bureaucrat' (Foilleux *et al.*, 2005: 617). Moreover, in their study, Foilleux *et al.* (2005: 617–8), argue that the Council and the EP possess different types of legitimacies (technical and democratic legitimacy respectively), but that the former 'wins' over the latter.[15] Interestingly, the Commission also seems to be affected by the Council's propensity to 'technicise' argumentation. According to Smith (2007: 232), the Commissioners adopt a more technocratic language when the official OLP has started in order to increase its chances of influencing the Council in the desired direction. 'However, they appear most free to act politically during the phase of European problem definition and agenda formation' (Smith, 2007: 232).

Initially, first reading agreements were 'legitimised' on the basis that they facilitated procedural efficiency when the issues at stake were only or highly technical and not politically contentious (see Farrell and Héritier, 2003a: 1197 and 2003b: 24). However, as we have seen, first reading agreements are today the rule and not the exception, as the majority of OLP dossiers are closed after the first reading.[16] It is consequently doubtful that it can be argued that they are all just technical and therefore politically uncontroversial.[17] Another issue is whether there is such a thing as a purely technical OLP dossier. Shackleton and Raunio (2003: 176) report the following:

> Certainly there was and remains a recognition, for example, that 'fast-track agreements' at first reading are a legitimate element of the legislative procedure. All agree that for technical dossiers, where neither side [i.e. the EP and the Council] is concerned to modify the Commission proposal, it makes eminent sense to avoid the establishment of a common position.

But how can the EP and the Council know that the dossier is only technical and consequently politically uncontroversial if it has not been subjected to public debate? Whether a matter is purely technical is not up to the Commission, the EP and the Council alone to define and decide. What is more, Rasmussen (2007: 5) also notes that dossiers are usually not either technical or political, but usually both at the same time. When seen in combination, the increased use of early agreements facilitated through secret trialogues and the limited number of actors (many of whom are unelected experts) as well as the procedural incentives to discuss in a technical rather than a politically explicit language, are all obstacles to an inclusionary will-formation process conducive to deliberative democracy.

To respect the normative hierarchy between representatives of affected and competent parties, then, is not only (even if mainly) about who is doing what (i.e. elected versus non-elected officials/experts), but also about in what form (in this case type of language) things are being done. To be a representative of affected parties involves making the latter aware of the normative dilemmas and trade-offs different solutions have. What is more, issues are never only technical and before they can be defined as such, they have to be 'identified, interpreted and specified in prior social processes. It has to be established through a "logic of discovery" that depends on a *perspective* and hence is normatively charged. In addition, application gives rise to normative concerns of its own' (Eriksen, 2009: 174). For citizens, this type of issue must be presented in an as clear and non-technical a manner as possible in order for the decision-makers to earn the label of 'representative' of affected parties. In this sense the ministers in the Council have obviously not earned the status of being representatives of their respective national citizenries. Even if this tendency is not as strong in the EP, the inclination to score points with the Council by adopting a technical language obviously does not strengthen the MEPs' status as the affected parties' representatives. That the MEPs buy into this 'ethically neutralising' path of how to make politics is arguably indicative of weak linkages between the EU decision-making system and the general publics.

Conclusion

In this chapter, two indicators were evaluated. The first concerned whether the main institutional decision-takers have a popular anchoring. Although the Commission is powerful, it is clear from Article 294 (TFEU) that a dossier cannot be adopted unless both the Council and the EP agree. Consequently, as both houses have the right to veto a proposal and have an electoral link, the first indicator is satisfactorily met.

The second indicator is more demanding and is targeted at finding out whether the organisation of actors who are actually participating in the moulding and shaping of a policy proposal respects the normative hierarchy between representatives of affected versus competent parties. As we have seen, the internal organisation in the Council does not comply with the principle behind this hierarchy as the unelected actors in the working parties, Secretariat and Coreper rather than the (indirectly) elected ministers dominate the Council's opinion- and will-formation processes. The fundamental democratic concern with the way the Council has organised its decision-making structure, highlighted in the 'A/B' point procedure where the ministers (usually) automatically accept 'A' points without any further ado, is that the ministers have actually abdicated from their role (and responsibility) as democratic decision-makers. The problem is that the role of democratic decision-maker cannot simply be handed over to officials and deputies without at the same time also degrading its normative status from *democratic* decision-maker to merely decision-maker. Hence,

> ...decision-makers may maintain that their [the working groups and Coreper's] solutions are correct or fair as far as they have managed to talk themselves into a consensus and agree upon the 'epistemic quality of the justification for political decisions'. This amounts merely to deliberation without democracy – to *technocratic deliberation* – as there is no chance for the affected citizens to say yes or no to the terms under which decisions are made.
>
> (Eriksen, 2009: 173)

In this regard, Rhinard (2002: 195–6) notes that 'This de facto decision-making authority is surprising in light of the fact that Council committees are not mentioned in the treaties and are only referred to once in the Council's institution-wide rules of procedure.'[18] The ministers are not 'hands-on', but depend too heavily on their officials. Moreover, being elected at one point in time does not give a *carte blanche* to speak on behalf of affected parties throughout the period between elections unless affected parties can also see/control what their representatives say and do on their behalf at regular intervals. From the perspective of deliberative democracy, the normatively superior position of being the spokesperson for affected parties' views is a *conditional status*, which the representatives have to 'earn' every time they claim to act on affected parties' behalf. This is not the case in the Council.

Within the EP, the situation is different. When seen in isolation, the division of labour between the committee and plenary settings makes good sense, and also from a democratic perspective, as the committee composition reflects the numerical strength of the plenary. In addition, the EP conducts its meetings in openness and has not transferred the opinion- and will-formation process to unelected officials as in the Council. However, when the internal organisation of the EP is viewed in the wider inter-institutional context of co-decision-making, the influence of the trialogues is seen to negatively affect the importance of parliamentary committees as the main place where decisions are discussed and prepared. This, obviously, also alters the legitimacy attached to the participation of directly elected MEPs. More specifically, the organisation of the EP, where the rules of inclusion and exclusion are first and foremost controlled by the political groups, becomes more difficult. The committee chairs, the group coordinators, the rapporteurs and, to some extent, the shadow rapporteurs all possess key positions in the decision-making process, they are 'legislative entrepreneurs' or 'specialist legislators' (Benedetto, 2005). Hence, when the key EP actors – first and foremost the rapporteur, committee chair, group coordinators – meet in trialogues before they meet in committee, the problem is that this also affects the internal power structure of the Parliament to the detriment of small party groups and independent MEPs. The use of trialogues seems to have increased the gap between big and small political groups as the importance of the committee setting has decreased. What is more, when deals are struck in trialogues and then presented to the EP, the parliamentary process of preference formation first in committee and subsequently in plenary sessions is constantly interrupted and upset. In sum, the trialogues largely disconnect the link between the deliberative opinion- and will-formation processes that take place in the EP committees and plenary sessions and thus reduce the democratic quality of the organisation of division of labour within the EP.

Even if the Commission is not a popularly elected body, it aspires to contribute to inclusion through its consultation system. On this basis, the Commission was assessed in order to see if the inclusion of the views of civil society organisations could somehow compensate for the lack of popular anchoring. The Commission is famous for its complex consultation system and includes a great variety of interests. However, it could not be argued that the system is capable of ensuring a democratic representation of societal positions. The reason for this is first and foremost because the Commission itself picks and chooses the types of interests that it finds more important. Moreover, to a considerable extent it also designs and shapes the types of interests that are included and involved (cf. co-optation argument). In addition, none of the groups or actors participating in the consultation system, and thus helping to influence the Commission's position, can themselves be held to account. In sum, these features of the organisation of the consultation system reduce the potential role external actors could play in helping to democratise the Commission.

The last issue discussed in this chapter pertains to the type of 'language' within which the OLP dossiers are framed and discussed, namely whether dossiers are primarily steeped in a political or in a technical type of language. As we have seen, there is a tendency that dossiers are discussed in a non-transparent and technical manner and that it is seemingly the lower levels (particularly the working groups) in the Council that set the standard in this regard. In order to win credibility and authority in the Council, MEPs are also adopting a more technical type of language. Hence, for citizens and other affected parties the political process becomes (even) more difficult to follow. When decision-makers dress up political issues in a technical language, the problem is that they implicitly also tend to exclude the possibility of forming a connection with the discussions taking place in the general public situated outside the institutionalised policy-making nexus. What is more, policy proposals are usually never just technical and it is not up to the decision-makers alone to define whether this is the case. Hence, when dossiers are not properly tested in an open and accessible manner, this only exacerbates the chances that the decisions that are reached will be deemed illegitimate, as they have not been properly 'tested'.

7 Openness and transparency

Introduction

The principle of openness and transparency is crucial for establishing a democratic dialogue between official political bodies and the wider public and for the latter to perform any kind of enlightened and informed critique and scrutiny of on-going legislative processes. A democratic decision-making procedure must consequently meet certain standards of transparency. From a deliberative perspective, which underlines an inclusionary and case-probing process prior to decision-making, it is vital to know not only what is on the agenda, but also where ideological positions diverge and dilemmas arise. In short, for citizens to develop an informed opinion about policy proposals, it is important that the decision-makers and their positions, as well as the process leading up to the adoption of dossiers are easily identifiable. In this sense openness is a crucial precondition for ensuring democratic decision-making. The EU has been heavily criticised for its remoteness and lack of openness in decision-making and has been attempting since the early 1990s to change that reputation through the introduction of various transparency measures. The task under this criterion is to assess whether decision-making under the OLP fulfils the institutional and procedural requirements for openness in the legislative process.

In this chapter, the OLP will be assessed against four indicators. The first reflects a minimum requirement of ensuring public access to the policy proposal before the final decision is taken. The second indicator is a continuation of the first, but is more demanding, namely that the relevant OLP documents shall be easily accessible from a digital register. The third indicator demands the transparency of policy-making debates whereas the fourth and last indicator requires that there is openness and intelligibility of the votes cast in OLP dossiers. As the trialogues and the Conciliation Committee are inter-institutional settings, and therefore not as such part of the individual institutions' set-up, the transparency of these two settings are consequently compared and assessed towards the end of the chapter.

Public access to policy proposals

> Indicator: all the involved institutional actors and the public can get *access* to the *policy proposal* before it is finally decided.

One minimum transparency requirement is that the policy proposal is available prior to final decision-making. The Commission's proposal is publicly accessible in the Official Journal once it is adopted in the College of Commissioners and sent to the EP and the Council. Hence, when formally launched, the document is easily obtainable from different sources such as the Commission's register of documents as well as in its Prelex service, the EP's Legislative Observatory (OEIL) and the EUR-Lex online service with regard to COM documents (which is the format in which Commission legislative proposals appear).[1] The Commission's Prelex service and the EP's OEIL service are both databases following the inter-institutional procedures/dossiers through the different decision-making stages. The Council does not offer a similar service as the EP and the Commission, but the Common Position is available from its register of documents.[2] The EUR-Lex service, on the other hand, is the EU's common (online) access point to EU law (containing documents published in the Official Journal, case law, Treaties, international agreements and parliamentary questions). Whereas access to OLP proposals is consequently easily available once the Commission has adopted a proposal and the formal procedure has started, access to the different versions of the draft proposal during the Commission preparatory process, is not publicly available. This fact notwithstanding, as OLP proposals are easily available during the official process through the Prelex, OEIL and EUR-Lex services, the first indicator of openness is complied with.

Access to documents

> Indicator: there is a digitally accessible register of documents (or other ways of ensuring easy access to documents).

Generally speaking, the OLP is the best-documented decision-making procedure in the EU. One reason for this is that access to decision-making acts is easier to achieve than for instance is the case with regard to decision-making in the EU's foreign and security policy (cf. Stie, 2010b). There is quite a large body of literature on the EU's policy on access to documents, both before and after the introduction of Article 15 (TFEU[3]) and Regulation 1049/2001.[4] Whereas a lot could be said, especially with regard to the exceptions to public disclosure that are allowed for in Article 4 and 9 of the Regulation, I shall not be dealing with this here. The reason is that the transparency situation with regard to the majority of policy areas falling under the OLP is not as often as other policy areas (for instance foreign and security policy) subject to these exception rules. Hence for OLP dossiers the situation is more straightforward and can be summed up as follows: both the Commission and the Council withhold preparatory documents, whereas the EP does not to the same extent. Second, the OLP is a 'prioritised' area for all three institutions in that there is comparatively much more information on this procedure than other policy-making procedures.[5] Having said this, the ongoing process of overhauling Regulation 1049/2001 has made many critics worried about the direction indicated by the EU institutions, especially the Council. At the time of writing, the Council has not yet reached a Common Position in the first reading, but the signals from the Council, according to Access Info (NGO), are not positive from a transparency perspective as it is, in reality, apparently suggesting stricter access rules on documents that it deems as 'requiring special protection'.[6] So, even if in the preliminary stage it could be argued that the Lisbon Treaty's Article 15 (TFEU) makes transparency and openness a more foundational or constitutional feature of the Union than the former Article 255, the jury is still out on whether this will actually be followed up in practice when the on-going process of re-evaluating Regulation 1049/2001 is finalised.

More specifically, according to Article 15 (TFEU: European Union, 2010) and further specified in Regulation 1049/2001(European Parliament and the Council of the European Union, 2001), the Commission, the EP and the Council are legally committed to make access to decision-making documents as publicly available as possible through digital registers (cf. Article 12(1) of the Regulation, see also Article 11). All three institutions have established such registers and thus ensure formal compliance with access to documents.[7]

All three registers are quite user-friendly and function well when one has a minimum of knowledge about the relevant document(s) one is searching for (such as number, title, key words, date etc.). However, if one would like to get a full overview of the relevant documents on a particular OLP dossier, the registers are of less help. In such cases, the aforementioned online services – Prelex and OEIL – present a much faster and easier path to a more complete picture of (on-going) OLP processes as they both provide chronological overviews of all publicly available OLP documents on a case-by-case basis.

In order to use the *Prelex* system, it is an advantage to know which document (series, number, year) one is looking for, but it is also possible to find a given dossier by using key words.[8] For every OLP dossier, Prelex provides a very good overview of when the proposal reaches the involved institutions. It also provides links to other relevant documents and lists the name of documents, their legal basis etc. The *OEIL* service provides a similar, chronological overview of ongoing dossiers. In other words, for ordinary citizens to get an overview of ongoing as well as completed co-decision-making, both services are invaluable channels of information and important supplements to the institutions' registers of documents.

It should be noted that the Council also offers an online service, which provides information on a dossier-by-dossier basis as well as by stage of the procedure.[9] There are, however, two specific drawbacks with this service. The first is that it is only available in French (or as it sometimes turns out when clicking on the different documents, in English). The second is that it is not as complete and thus not as user-friendly as the Prelex and OEIL services. In addition and as opposed to Prelex and OEIL, which both contain all publicly available documents from all institutions, the Council service only contains Council official documents, neither drafts nor lower level working documents (including documents from the working parties).

With regard to the institutions' registers of documents, the *Commission's* register contains C, COM, SEC, agendas (OJ), minutes (PV) of Commission meetings, Commission staff working papers (SWD) and Commission and High Representative Joint Acts. It is the four latter categories that are interesting from an OLP perspective. Whereas COM documents represent proposed legislation, SEC documents are 'Documents which cannot be classified in any of the other series'.[10] Both the SWD and the SEC categories are obviously interesting, but there are seldom publicly disclosed preparatory documents available apart from impact assessments, which also appear on the EUR-Lex online service on a regular basis.[11] An obvious drawback/shortcoming with regard to the functioning of the Commission register is that for each document, interested citizens must fill out and send an online request form to the Commission who will then consider the matter. In other words, many of the documents are not directly retrievable from the website. What is more, other preparatory documents such as online consultation documents and Commission expert group documents, as well as the DGs' documents are, if at all publicly available, not posted on the register. Altogether this obviously makes public access to Commission documents cumbersome.

As can be anticipated on the basis of the foregoing chapters, access to EP documents is readily available in the register or at the websites of the various committees. These cover all EP documents relating to plenary sessions, committees, delegations and individual members, that is, all relevant OLP documents originating from the EP such as minutes, amendments, agendas etc. Conciliation documents are only available on an ex post basis in order not to disadvantage the EP's bargaining position by disclosing the Parliament's flexibility/limits towards the Council in the Conciliation Committee (see also rule 104, EP rules of procedure). According to the EP's annual report on access to documents 2010, 90 per cent of all the documents in the register are directly available from the internet.[12] Those documents that are not directly available to the public concern first and foremost correspondence and comitology procedures (ibid., p. 9).[13] Hence with regard to OLP documents, the register provides satisfactory access.

Whereas there are hardly any restrictions with regard to accessibility of EP OLP documents, the Council[14] – as the Commission – is generally a more secretive institution. According to Annex II in the Council's rules of procedure, the following documents relevant for co-decision-making are available in the register:

1 *Provisional agendas* for Council meetings and for its preparatory bodies.
2 Documents submitted to the Council, which are listed under an item on its

agenda marked with the words *'public deliberation'* or *'public debate'* in accordance with Article 8 of the rules of procedure.

3 In the legislative field, 'I/A' and 'A' item notes submitted to Coreper and/or the Council, as well as draft legislative acts, draft common positions and joint texts approved by the Conciliation Committee to which they refer.
4 Acts adopted by the Council during an ordinary (or special) legislative procedure and joint texts approved by the Conciliation Committee under the OLP.
5 Any other text adopted by the Council which is intended for publication in the Official Journal.
6 Documents originating from a third-party, which have been made public by the author or with his agreement.
7 Documents which have been made available in full to a member of the public who made an application.

Whereas this seems like a comprehensive list, it should be noted that there are no preparatory documents available apart from the draft agendas. However, as we have seen in Chapter 5, most decisions are taken in the working parties and Coreper so, apart from provisional agendas, there are in reality no documents available in the Council's register (or in any other database) from these meetings. Hence, even if the Council is at its most open when dealing with OLP dossiers, there are nevertheless restrictions on the availability of documents, which represents a shortcoming from a democratic point of view.

A final note concerning translation should be made. The many official languages of the EU present an obvious challenge, namely to provide full transaltions of them all, and at every meeting. This task has been made more difficult by the last enlargement. Documents accessed from the EUR-Lex database are available in all the official EU languages. Hence all the official versions of legislative (OLP) proposals – the Commission proposal, the EP's Position, the Council's Common Position, the Joint Text resulting from the Conciliation Committee – are translated into all the EU languages. This is not the case with all documents in the Prelex and OEIL services which are often only available in English and French (see also Hayes-Renshaw and Wallace, 2003: 39–40).

Openness of debates

> Indicator: there are either verbatim records, video streaming of debates through the internet or minutes available in all official languages.

Through the assessment of meeting places, we already have an idea of how openness under co-decision-making patterns and that the transparency conditions differ depending on the institutional actor. Unsurprisingly, the EP is the most open.

Since 1999, both the *EP's* plenary sessions as well as the committee meetings have been open. As we have seen, *plenary* sessions are very easily accessible through video streaming available on the EP's internet site. This multimedia library service provides an archive of already conducted sessions (although it is quite meagre, since it has only been operative from July 2006) as well as the possibility of following live debates as they take place. It is easy to manoeuvre one's way around the archive as it not only provides viewers with the opportunity to see and hear the MEPs discuss, but there is also an overview of topics (subdivided into debate or vote) and speakers. This makes it easy to find one's way around a meeting and possibly skip parts that are not of interest. When a MEP is speaking, his/her full name, party affiliation, nationality and current subject are presented. In addition, there is a translation access point where the audience can choose any one of the official languages. In sum, this service is and will become, an even more valuable source of information over the years as the archive is filled up. Live and stored video transmissions of meetings in addition to minutes[15] and electronic verbatim records in all the EU's official languages consequently makes the openness of EP plenary debates very good and obviously contributes to more transparency of EU decision-making.

Also, the EP committees are broadcast live and it is now possible to follow them online. Consequently, when we compare the transparency situation in the Parliament to the former parliamentary period, the overall accessibility and openness of the EP has improved considerably, especially when we take into consideration how important committee meetings are for the legislative process. It should, however, be noted that translations in all official languages are not available from committee meetings. Hence, in addition to the lack of verbatim records, openness improvements are still needed.

Whereas the conditionality rule allowing for closed plenary settings in exceptional cases was abandoned in the 1999 revision of the rules of procedure, the EP *committee* meetings can be closed under three specific circumstances:

> First, committees may decide to discuss certain designated matters behind closed doors as long as such a decision is taken at the latest when the agenda of the meeting is adopted. Second, discussions on request to waive a member's immunity must be held in camera. Third, the conciliation committee and the Parliament delegation meetings that precede the committee are not public.
>
> (Corbett *et al.*, 2007: 334, see also pp. 168–9)

In short, this means that EP committee meetings as such are hardly ever closed under the OLP.

There is also another possibility of getting to know more about ongoing processes. Under the 'Conciliation and Codecision' internet site,[16] the Parliament provides more information about the progress of OLP dossiers. Under the access point 'ongoing procedures',[17] there is a list of all dossiers currently under treatment in the Parliament. The list contains information about the responsible EP committee, name and political group of the rapporteur, title and number of the dossier as well as a column indicating the likelihood that there will be an early agreement/first reading agreement on the dossier.

In addition, each committee has their own website where they present information about their work and collect documents which are accessible for everyone who may be interested. Here, a calendar with dates and agendas for upcoming meetings, minutes from meetings as well as committee reports can be downloaded. The websites contain different types of documents ranging from Commission proposals, draft opinions, draft reports and working documents, to minutes and agendas.[18] As far as meeting documents are concerned, these are presented in all the official languages. Moreover, all the committees provide additional information about their work and some, such as the Committee on Environment, Public Health and Food Safety[19] and the Committee on Regional Development,[20] have their own newsletter and references to relevant external links. Most committees also conduct hearings. The documents from these meetings are usually available from the websites of each individual committee. These documents contain prepared statements, agendas and background reports. In sum, Corbett *et al.* (2007: 334) argue that this makes the EP on a par with, if not better than, most national parliaments – even the Finnish, Swedish and Danish parliaments cannot compete with the EP in this regard as their committee meetings are mostly closed.

As argued earlier, the secrecy of both EP delegation meetings and the Conciliation Committee is justified because it contributes to ensure that the EP is not disadvantaged in its negotiations with the Council and because the documents are disclosed on an ex post basis (I come back to the Conciliation Committee below).

Whereas the general rule with regard to the EP is that it conducts all its committee and plenary meetings in public, the Council and to some degree also the Commission, do not.

According to the *Commission's* rules of procedure, the meetings of the Commission shall be confidential (Article 9) and this provision does not only apply to the College of Commissioners. Apart from open consultations and to some extent the dialogue and policy forums, the Commission OLP settings are closed to the general public. There are no publicly accessible verbatim records or internet streaming services available. The principle of collegiality commits the Commission as a collective to defend all decisions issued in its name (see Nugent, 2001: 92) and thus guard its internal debates.

With regard to the *Council*, the implementation of the access provisions is found primarily in Annex II of the rules of procedure. To open up the Council decision-making processes has been and still is a recurrent theme on the EU agenda. We have seen that it is now committed to more openness when dealing with OLP dossiers. Whereas the meetings of Coreper and the working parties are closed both with regard to verbatim records/minutes and video streaming, Council ministerial meetings are now required to be open when the ministers deliberate and vote on legislative drafts.[21] This means that it is not only possible to follow the adoption of 'A' points, but also more substantial discussions on 'B' points.

As we have seen in Chapter 5, the Council provides a video streaming service similar to that of the EP, where it is possible to follow some debates live either on the internet or in an overflow room in the Council building. The openness provisions on OLP dossiers are obviously much better in comparison to other decision-making procedures in the EU, and especially so since it is required that the ministerial meetings are open when 'B' points are discussed. We have, however, also seen that the number of 'A' points has slightly increased and the number of 'B' points has decreased significantly (cf. Settembri, 2007). In addition, enlargement has made recourse to more informal meetings outside the public eye more common. This obviously reduces the importance of the video transmitted policy debates.

What is more, the lack of openness in the deliberations of Coreper and the working parties cannot be ignored when most of Council decision-making passes as 'A' points without any discussion at the ministerial level. As we have seen, Hayes-Renshaw and Wallace (2006: 52) estimate that as much as 85–90 per cent of all Council decisions are taken at the lower levels, 70 per cent on working party level and 10–15 per cent at the level of Coreper. The lack of openness in these settings is not only a serious problem for the citizens, it is also a major problem for the EP. Unlike the Commission, which has access to all Council settings, the Parliament is excluded from these arenas and MEPs must inform themselves about the progress in the Council either through the updates on the websites (like ordinary citizens) or through personal contacts within the Council. MEPs participating in trialogues are obviously in a better position as they are constantly in contact with both Commission and Council representatives. In sum, we may say that given that the currency of influence is very much dependent upon information, the lack of inclusion reported under the second criterion is intimately tied to the non-transparent nature of the preparatory bodies of the Council (see also Menéndez, 2009: 302). Moreover, the lack of openness in the Council systematically favours competent parties to the detriment of affected parties since the majority of MEPs are unable to properly follow the preparatory processes in the Council.[22] This is not the case for Council ministers who, as we have seen, often have their officials positioned in the EP as listening posts throughout the legislative process.

Another openness aspect pertains to the rules on how to present/report legislative amendments. If we compare the EP and the Council, there is a striking difference between the two institutions with regard to transparency. In the EP, amendments must be made in writing indicating the place in the text where deletion and/or insertion of amendments should be made. The proposal must be accompanied by a written justification and the full name of the elected politician filing the amendment(s). This is not the case in the Council where there is no indication in the text (underlined in italics, bold or the like) with regard to where new, modified or deleted amendments can be found.

In comparison with the EP committees and the amount of relevant information that can be found on each committee's website, the internet information on each Council configuration is quite limited. When accessing the link to one of the configurations it is possible to find a short description of what it does, that is, in which policy areas it is involved. Each configuration website provides a link to its press releases dating back to approximately 1999 when the Council register was first introduced. In addition, summaries of previous Council meetings, which are rewritten from the minutes available under the aforementioned 'documents' access point, can also be found.

There are many dead ends when trying to recapture what took place in Council meetings and outside actors must have patience and time to figure out where relevant information is stored (if it is openly available) together with knowledge about how the system works. In sum, whereas the treatment of co-decision dossiers represents the Council at its most open, it still does not meet the democratic requirement of transparency of debates.

Intelligibility of voting

> Indicator: there are available voting records which include information and explanation about who voted and what position they defended.

A public decision-making process must end in an outcome – either a proposal is rejected or accepted – and this should also be as transparent as possible. Generally speaking, EU decision-makers are, according to Article 296 (TFEU), obliged to provide reasons for the results they reach: 'Legal acts shall state the reasons on which they are based and shall refer to any proposals, initiatives, recommendations, requests or opinions required by the Treaties.' As we can see, this also includes OLP acts. But whereas Article 296 refers to the final result, it does not tell us anything about how the institutions voted individually. An important part of democratic decision-making is to make publicly visible and easily accessible the results of how representatives voted. Public display of voting results is an effective and undemanding mechanism that provides the citizens with the possibility to make an informed opinion about their representatives and to hold them to account in the next election: 'A vote is in effect the manifestation of the responsibility of an individual (representing different interests, depending on the institution of affiliation) with regard to a decision' (Settembri, 2005: 642). Let us therefore look at the extent to which the institutions ensure intelligibility of votes.

As with the previous indicator, intelligibility of voting in the *Commission* is also affected by the principle of collegiality. Potential disagreements among the Commissioners remain internal as all decisions are expected to be defended collectively. The voting result on whether to launch an OLP proposal is therefore always unanimous. However, it is more important that the EP and the Council provide a public display of voting results, as they are the final decision-makers with veto power.

According to the *EP's* rules of procedure (rules 165–169),[23] there are basically four ways of voting in the EP, either by a (1) show of hands, (2) electronically, (3) secret ballot[24] or by (4) roll calls.[25] From a transparency perspective, it is the latter that offers the best accountability protection, as this is the only time where individual MEPs' votes are formally registered:

> first in the annex to the minutes which appears the next day, and later in the translated minutes which come out in the Official Journal about three months later (although unofficially a print-out is available in the press room within half an hour of the vote).
>
> (Corbett *et al.*, 2007: 175)

Roll calls are, however, not the most widely used system, as they must be requested on a case-by-case basis and require the support of one of the political groups or at least 40 MEPs before the vote is to take place. Settembri (2005: 647) estimates that only one-quarter to one-third of the total number of votes taken in the EP are subject to roll calls and that this result is not sufficient for a democratic monitoring of votes in the EP.[26]

In most cases, votes are taken by a show of hands, which gives no possibility of tracing who voted for what as it only shows how many supported and how many rejected the proposal. When votes are taken in this way, the group coordinators point their thumbs up or down to indicate to fellow members the political group's position on a specific vote. Electronic voting is used in cases where the EP President is uncertain about the vote and shows how many voted yes and no as well as the number of abstentions. Also, in this case, personal votes are not displayed.

With regard to explanations of votes, the EP seems to have the opposite problem of what we are addressing here, as rule 170 of the Parliament's rules of procedure rather lists the restrictions on how and when it is possible to state the reasons for the votes cast.

So far, only voting in plenary sessions has been assessed. There are two ways of voting in committee, by the show of hands or by roll call votes (see rule 195, EP rules of procedure), although the latter is 'extremely rare' (Settembri, 2005: 648). Given that the openness of committee debates is not as good as that of plenary sessions, committee voting suffers and is consequently also less transparent. This also affects voting in plenary debates as:

> the only information reported in the minutes is whether the committee has approved an amendment or not, and there is no chance of finding out who actually supported or opposed it (unless the vote was unanimous). The procedural page of all reports gives only an indication of the result and does not say how each member voted.
>
> (Settembri, 2005: 648)

Altogether, even if it is not impossible to find out how individual MEPs voted in committee and in plenary, the Parliament's system for reporting on the votes that have been cast is not satisfactory. Settembri (2005: 650–1) sums up the EP voting system in the following way:

> the EP voting is neither entirely secret nor fully open, but somewhere in between. Where voting is not explicitly secret ... or deliberately fully transparent (by roll-call) ..., voting is public by name only, because the voting system excludes any possibility of individual accountability. Contrary to the practice in most national parliaments, the EP works on the basis of unrecorded deliberations with fully transparent votes taking place only where explicitly requested.

In addition, if we take into consideration that the debate and the casting of votes are separated in time, i.e. that a vote does not follow directly after a debate, it can be very hard to get an overview of the situation. In order to provide full intelligibility of voting, the EP should go for roll-call votes on every dossier both in committee and in plenary sessions as this gives citizens the possibility to readily achieve an overview of the MEPs individual voting patterns.

According to its rules of procedure (Article 7), the results and explanations of votes taken in the *Council* during legislative procedures shall be public.[27] This covers both the adoption of the Common Position taken during the procedure, the final outcome of a legislative act and the results of votes taken in the Conciliation Committee (Bauer, 2004: 373). Again, transparency of Council procedures is better when it is acting in its legislative – as it does under the OLP – rather than in its executive role as it does under for instance in foreign, security and defence policy.

We saw that the EP votes mostly by raising hands and only occasionally by full roll-call votes. In the Council, the voting system is arguably even more implicit (due to the secrecy of meetings prior to the dossiers reaching the ministers and the propensity to act by consensus even where a qualified majority applies) even if roll calls do occur. However, with the establishment of the register (in 1999), the formal reporting of votes has become better and, unlike the EP, information is now given on who abstained and/or voted against an act, that is, on those acts where votes were explicitly contested. The voting chart attached to the Council minutes or in the monthly summaries[28] of legislative acts shows the results of votes, who voted for and against, and who abstained. As we have seen, in these documents, the member states, the Commission or the Council can attach statements explaining what in their view is the weak part or the problem with the outcome of the vote, but attaching such statements is not obligatory. In addition, the Secretariat publishes a press release from each Council session.[29] In other words, beyond the information of who voted what, the explanation of votes is voluntary and appears when the member states or Commission want to make a statement. As we have seen, Hagemann and De Clerck-Sachsse (2007) found that since enlargement member states have been using this opportunity more often and it is arguably a way of stating disagreement without blocking the vote. Mattila (2004) contends that when the member states do not find it worthwhile – either because the issues are not very controversial and/or they will not get valuable media attention in the national public sphere – they will abstain from reporting dissent. Moreover, 'In a system that requires agreed decisions to be implemented in domestic law, consensus encourages compliance, and an outvoted government might evade this' (Hayes-Renshaw and Wallace, 2006: 278; see also Hagemann and De Clerck-Sachsse, 2007: 20).

Another point, and similar to the situation in the EP, is that Council voting procedures are also separated from the debates on the relevant OLP dossiers (Mattila, 2004: 48; see also Bauer, 2004). In other words, 'A' points in cases belonging to a different policy area may be decided by a different Council configuration in order for a decision not to have to wait until the next meeting of the responsible Council configuration.[30]

In sum, intelligibility of voting is not complete in either of the institutions. The Council is doing well in that it provides a voting chart of how each member state voted, but explanation/justification of the votes is optional. Even if the votes are public in the EP, the intelligibility of voting suffers seriously as roll calls are not the mechanism most commonly used. Finally, the Commission's principle of collegiality effectively precludes the possibility of holding individual Commissioners to account.

In this regard, critics will perhaps argue that the Commission should not be evaluated according to the same openness standards as the EP and the Council, as the Commission is 'an institution that wields its responsibility collectively, control over internal debates and the votes of the individual commissioners would be inappropriate' (Settembri, 2005: 644). Settembri does not, however, provide us with the reasons for what constitutes appropriate and inappropriate standards for Commissioners' policy-framing powers. This is most likely due to the indeterminate status pertaining to the kind of institution the Commission is and should be. In other words, is it (or should it be) conceived as a type or supranational version of 'government' or is it (or should it be) conceived as a supranational executive administration? In national contexts, it is common for both an elected government and the administration/bureaucracy to act with one voice. Hence internal deliberations may be shielded both during and after session, thus ending up with a decision (and a justification) that everyone is committed to defend (even when they are disagreeing). The difference between the national and the European levels is, however, that in the former the governments and bureaucracies form part of larger constitutionally settled accountability and responsibility structures. Hence the roles of governments and bureaucracies are more clearly defined and are subjected to better democratic control. In the EU, however, the Commission is appointed by the member states, and is not designed to represent the national, but rather the European interest. In this regard, the question arises of whether the Commission is legitimately authorised to speak on behalf of the European interest as long as it is not democratically elected. What is more, the Commission is exercising both political and administrative roles and is in this sense (and from a conventional perspective) a 'bastard' institution in disrupting the principle of separation of powers.

The trialogue versus the Conciliation Committee

So far the inter-institutional settings – trialogues and the Conciliation Committee – have only been referred to indirectly. Both the trialogues and the Conciliation Committee are closed forums, but unlike the former, the transparency rules are stricter with regard to the Conciliation Committee in that information is provided before and after such meetings have taken place. Moreover, the difference between them has very much to do with their different levels of formality. As we have seen, the mandate, rules, time limits, stages, participating institutions, rights and duties regulating the work of the Conciliation Committee are set out in the Treaties and further elaborated upon in the rules of procedures of the three institutions and also in the Joint Declarations. They are, in other words, well known and easily accessible.

The trialogues (which are not mentioned in the Treaties) lived a rather anonymous life until the first Joint Declaration between the institutions on the practical arrangements for the OLP was introduced in 1999 and then further elaborated upon in the 2007 revised version. With these declarations, we may say that trialogues were semi-formalised and given a set of standard operating procedures pertaining to a commitment by all three institutions to conduct trialogues throughout the procedure. The strange thing, however, is that whilst the Joint Declaration from 2007 underlines a commitment to the principles of transparency, accountability and efficiency (paragraph 3), the whole set-up of trialogues violates the first two principles. The very purpose with the trialogues is that they are conducted behind closed doors, there are no drafts or documents available prior to or after the meetings and there is obviously no openness of debate. Since trialogues are not official decision-making settings no votes are taken. Trialogues do not further transparency and accountability, only efficiency (see also Farrell and Héritier, 2003b: 9).

Peers (2006) particularly criticises first reading trialogues for their lack of 'information to the public as to whether first-reading discussions are even taking place, and no information as to the drafts under discussion.' Peers points to the difference between first and second reading trialogues and conciliation trialogues, finding the latter two more legitimate. The reason is that they at least operate under a semi-formal type of framework 'which entails public knowledge about the state of the procedure and the texts under discussion, as well as a public explanation of those texts (for example, the Council's statement of reasons for adopting a common position' (ibid). An (ironic) illustration of the problem of transparency pertains to the treatment of Regulation 1049/2001, which covers the substantiation of former Article 15 (TFEU), i.e. Article 255 (TEC) on public access to documents. In an open letter directed to the EP (Statewatch, 2001a), the NGOs particularly criticised the extensive use of trialogues. They argued that the Regulation:

> has been adopted as a result of 'trilogue' negotiations with the Council (and the European Commission) which have taken place behind closed doors for over five months. At no stage has a full debate in the parliament taken place on the various substantive issues proposed. We believe that the procedure followed is not only inappropriate given the nature of the topic in question, citizens access to information, but also substantially weakens the nature and purpose of the co-decision procedure as such and parliament's function in that respect.[31]

In other words, the problem is that outside actors do not know which issues are on (and which do not enter) the negotiating table in first reading trialogues. Moreover:

> If the two institutions wish to conclude at first reading, without a common position being adopted, Council has to be able to convince Parliament to adopt in plenary amendments that it (Council) can accept. In other words, it has to intervene in the process by which Parliament determines its first reading, without itself having established its position, and thus both institutions are negotiating without either having finalized a point of departure for negotiation.
>
> (Shackleton and Raunio, 2003: 178)

Even if closed, the meetings in the Conciliation Committee are, in this sense, a very different matter as 'the context of negotiations is very clearly defined: Parliament has adopted amendments to the common position by absolute majority … and the Council is obliged to respond to the amendments, accepting them, rejecting them or proposing a compromise' (Shackleton and Raunio, 2003: 178). We also know when the meetings are held and who the participants are, and documents are made available after sessions are completed.[32] In sum, the above conditions under which the Conciliation Committee works, make the secrecy of its meetings (more) legitimate as there is a certain ex post transparency (and thus also accountability) revealing what has been debated once decisions have been reached. What is more, as discussed in Chapter 5, the Conciliation Committee is set up as a final bargaining setting when the preceding stages have been exhausted, but when the parties are still motivated to conciliate. Hence, the various positions are by then known and familiar.

The same cannot be said about the trialogues. Trialogues take place at all stages of the procedure, but the meetings are not publicly announced. There are no meeting records hence the identity of the participants can only be known indirectly, in the sense that it is usually the same actors who are involved (i.e. not individuals, but by virtue of position: Coreper, EP rapporteur, chair and Commission official). There are also no documents available from these meetings either during or after they are held. Hence for the public, the discussions are completely sealed off.

Conclusion

In this chapter, I have assessed the OLP against four transparency indicators. The first is a minimum and low-threshold indicator, namely that the policy proposal should be publicly available prior to final decision-taking. Even if the proposal is not generally available during the Commission's preparatory phase, it is published on the EUR-Lex service the moment the official OLP starts and the proposal is formally launched. Hence, the OLP has no problem in complying with this requirement.

The second indicator is more demanding but nevertheless is, in principle, quite easy to meet in the age of internet and computer facilities. A good register is characterised by clarity, user-friendliness and also that it provides an exhaustive impression, i.e. the source where all the relevant documents can be found (cf. Curtin, 2000, 2003). Ideally, there should be a centrally organised register rather than every institutional actor providing their own register, but the latter is also satisfactory. As we have seen, there is not one single and centralised register covering all three institutions where all the documents can be accessed.[33] Each institution has its own system, concerning both how documents are archived and the way they are presented. Hence the overall framework for accessing documents through the registers is quite complex and can be confusing and time-consuming for citizens to find their way around. From a democratic point of view, a central register would make it easier for citizens to get an overview of Union documents.

The lack of a centralised register notwithstanding, the institutions' registers and the Prelex and OEIL services together offer a good route to the *publicly* available OLP documents. The good news, then, is that in comparison to (for example) documents concerning the EU's foreign and security policy (see Stie, 2010b), access to OLP documents is quite good. More specifically, whereas the availability of EP documents 'scores' well, the lack of access to working parties and Coreper documents in the Council is not satisfactory given that so many decisions are actually taken there. In addition, it is difficult to get an exhaustive overview of the relevant documents during the Commission preparatory phase. Overall, access to documents under the OLP is not fully satisfactory.

The third indicator is concerned with the openness of debates. In short, the Commission and the Council fail to meet this demand, whereas the EP succeeds even if the transparency of committee meetings could be better. The meetings of the Commission are bound by the principle of confidentiality and thus explain the secrecy of meetings. The Council under the OLP, on the other hand, are more open than under any other procedural arrangements. When it fails to meet the demand of transparency of debates it is because the major chunk of legislative discussions are not conducted in open ministerial meetings, but in the lower echelons of the Council hierarchy where the public is still excluded. In addition, Council ministerial meetings are increasingly accompanied by informal meetings conducted outside the public eye. Again, there is a mixed result with regard to whether the OLP meets the requirements of transparency of debates.

The fourth and final indicator looked at the intelligibility of voting. Whereas the Commission's principle of collegiality commits the Commissioners to unanimity and thus renders voting in this institution uninteresting, the Council provides a very useful voting chart attached to the Common Position. Here it is possible to find out who voted for and against and who abstained. There is, however, no duty to state the reasons behind individual votes and Council member states arguably only decide to attach a statement when political gains can be achieved. This reduces somewhat the intelligibility of votes in the Council, but this is far from detrimental and since the chart provides easy access to who voted what, the Council clearly passes the formal requirement of intelligibility of voting. The EP also meets this requirement though with reservations, due to the difficulty of tracking down who voted what. Had roll-call voting been the default procedure (instead of just reporting the number of for, against and abstained votes resulting from a show of hands and/or electronic voting) there would have been full intelligibility of voting in Parliament. However, as the discussions are open in the EP it is not impossible to know who voted what, though it is very cumbersome to keep track of what every MEP voted individually.

Finally, the openness situation in the trialogues and Conciliation Committee was also evaluated. Whereas the former fails in regard to all openness requirements (no access to documents or debates, no information of meetings or participants), the latter does not. The reason is that even if the Conciliation Committee is closed during sessions, the working documents are available after the meetings are over and this makes it possible for citizens to scrutinise the proceedings ex post. Moreover, the Conciliation Committee is only convened towards the very end of the process when the positions are well known and where no agreement has yet been reached, but where there is still a possibility for this to happen. There is also information about who the participants are, and when the meetings are going to be held. Hence, in contrast to trialogues, the Conciliation Committee provides an example of when non-public discussions are legitimated from a democratic point of view (cf. Warren, 2007: 282).

8 Neutralisation of asymmetrical power relations

Introduction

One of the main tasks of a democratic polity is to provide institutional arrangements that, ideally, are able to transform, but in practice are, at least to some extent, capable of neutralising and/or altering the parameters of power politics, social inequalities and strategic action. As mentioned earlier, here I distinguish between two broad categories of neutralisation mechanisms. The first covers mechanisms that are internal to the procedure and detectable in the procedural set-up, i.e. ex ante mechanisms. The second category covers complaint procedures that are external mechanisms kicking in after the decision-making process is over, i.e. ex post accountability mechanisms. In this chapter, I look at two ex ante mechanisms: Intelligibility of procedure and separation of powers. In addition, I investigate two ex post mechanisms: judicial review and Ombudsman.

Intelligibility of procedure

> Indicator: there is a *precise description of the rules* governing co-decision-making. The *Treaties* contain information about the main actors and what powers, rights and duties they have. In addition, there is information about the decision-making rule (unanimity/majority vote) and the sequencing of policy stages. The institutions' *rules of procedure and/or inter-institutional agreements* are in line with Treaty specifications.

There is power in knowing the rules of the game, hence to better the information asymmetry between those involved in decision-making and those affected by it, clear and unambiguous rules – or what is called 'intelligibility of procedure' in this book – help to neutralise such inequalities. This is always easier when the rules are written down, when the procedural steps are predictable and consistent and when informal practices do not deviate too much from the formal set-up. However, formal rules are usually not precise enough to avoid the development of informal practices, though the extent to which these provisions are precisely elaborated and written down can at least contribute towards ensuring that informal practices are not totally incompatible with the intention of formal regulations. Moreover, this can also empower those affected and/or the media in the sense that if informal practices clearly deviate from formal provisions, they can (try to) hold decision-makers to account by pointing out the discrepancy and by claiming justification and/or revision of the rules.

Precise and sufficient description of OLP rules?

The OLP is described in Article 294 (TFEU). In contrast to the Nice provisions, the adoption of the Lisbon Treaty and the renaming of the co-decision procedure as the ordinary legislative procedure have established it as *the* legislative procedure of the EU. This has also helped to make it more visible within the Treaty framework (cf. Stie, 2010a). Earlier, Union citizens needed to know which Treaty article to look for in order to get an overview of the basic content of the procedure. However, even if this was the case, the term – co-decision – was used everywhere else but in the Treaties, e.g. in formal documents originating from the official EU institutions and on the Union's websites.

Article 294, which describes the basic rules of the procedure, states that the institutional actors eligible for participation are the Commission, the EP and the Council. It indicates that the Commission has the right to initiate policy proposals and otherwise function as a facilitator, whereas the EP and the Council serve as co-legislators, both having veto power on the final outcome. Moreover, it provides detailed information about decision-making rules and time limits in the EP and the Council under the second and third readings (the first reading has no time limits) and it describes the basic format of the Conciliation Committee (purpose, participants, voting rules, etc.). In addition, it states how a proposal can be adopted after each reading.

As mentioned above, with the coming into force of the Lisbon Treaty and the renaming of the procedure, the policy areas subject to legislative treatment according to the OLP have doubled and now cover the vast majority of legislation in the EU. Earlier, when the OLP was not the default procedure, it was not as straightforward as it is now to get an overview of the areas where co-decision and (those where) consultation, cooperation and assent procedures applied respectively. This information is still scattered around the Treaties, but since the majority of policy areas are now regulated under the OLP provisions, it is at least not unreasonable to (more or less automatically) assume that a policy proposal is to be treated according to the OLP. However, the easiest way to find out which policy areas are covered by the OLP is to consult the EU website.[1] In addition, different decision-making rules apply both with regard to the Council as well as the EP. The Council usually decides (at least formally) by QMV. But, in cases where the Council sides with the EP, and the Commission is in disagreement, the Council can only uphold the EP amendments if it is able to drum up unanimity. With regard to the EP, simple majority (both in committee and plenary) is required for the adoption of the first reading Position as well as for the Joint Text in the third reading, whereas absolute majority is needed in the second reading.

The institutions have themselves done quite a lot to ease the apparent obstacle of getting acquainted with the formalities of the OLP on their respective websites. From the point of view of this indicator, all three institutions have provided satisfactory information on how the procedure works through their own 'step-by-step' introductions or guides to the OLP.[2] In addition to the institutions' rules of procedure, it can also be argued that these documents together offer detailed and helpful information on the formalities and workings of the procedure. Hence the OLP is thoroughly explained and well documented.

It may also be argued that Article 294 is fairly comprehensible and thus describes the different stages in the OLP process satisfactorily, i.e. the possibility for three readings, potential recourse to the Conciliation Committee, which actors are formally involved, voting procedures etc. Maurer (2003: 244) also points out that the procedure is not outstandingly difficult as it 'is as cumbersome as the joint decision between the German Bundestag and the Bundesrat or between the French National Assembly and the Senate'. In other words, the OLP is not necessarily more complicated than many national decision-making methods. It has, however, been argued that the OLP 'is a complex procedure whose unwritten rules and behavioural norms have developed, are still developing and are not easy to quickly grasp. If all EU negotiating is a dark mystery, legislative co-decision is a blacker shade of dark' (Bostock, 2002: 222; see also Häge and Kaeding, 2007; Fouilleux *et al.*, 2005). The informal rules and practices mentioned in the citation first and foremost allude to the trialogues as well as the aforementioned grand coalition trend between EPP and S&D in the Parliament (Andeweg, 2007: 108). The problem is that these rules and practices tend to depoliticise OLP politics as both aspects contribute towards hiding political salient discussions outside the public radar.

To most Union citizens, the EU is often perceived to be a remote and opaque political system in the sense that it is hard to orient oneself and keep track of the different rules according to which policies are adopted in the different sectors. Even if it can still be confusing and difficult to immediately distinguish (the procedural difference and impact) between legislative, non-legislative and budgetary decisions, the Lisbon provisions and the renaming of co-decision to the OLP has at least simplified the Union's legislative activities and enabled them to be more readily understood; it has thus contributed to more systemic transparency.

Correspondence between Treaties versus Joint Declarations and rules of procedure

The trialogues are, however, not described in the Treaties, but rather in the two inter-institutional agreements specifying the goals and rules regulating them.[3] Whereas the first Joint Declaration is more focused on the conciliation process, the second agreement is devoted to trialogues in the first reading. We may say that these two agreements contribute to semi-formalise the trialogues, and thus somewhat alter the character of the OLP as it is described in the Treaties. As we have seen, the formal design is sequential or (more accurately) described as a relay between the institutions, but in reality, and as described in the Joint Declarations, it is now more of an interlinked and continuous process, where a limited number of participants from the EP and the Council are in constant contact with each other throughout the procedure (Shackleton and Raunio, 2003: 173; Farrell and Héritier, 2004).

On this basis, it can be argued that the actual practice of the OLP as well as the Joint Declarations deviates quite substantially from the Treaty prescriptions and consequently obscures the fairly clear *formal* set-up of the procedure. In fact, Farrell and Héritier (2003b) argue that the trialogues have led to 'an invisible transformation' of the OLP. What is more, Settembri (2007: 32) even goes as far as arguing that the inter-institutional relations have been perverted:

> The high frequency of codecision files concluded at 1st reading, the fact that emissaries of Council and Parliament get together to broker an agreement since the outset of the procedure, the practice of adopting package deals amalgamating amendments from both institutions suggest that a new type of procedure is surfacing.

In sum, the intelligibility of the formal OLP is generally quite good, but the discrepancy between Treaty provisions, on the one hand, and inter-institutional agreements and actual practice, on the other, obviously weakens the OLP's fulfilment of this indicator; the extended use of trialogues also exacerbates the complexity of the procedure and thus its comprehensibility for the general public (I come back to this more closely in the next chapter).

Separation of powers

> Indicator: the institutional actor presenting a policy proposal is not at the same time the decision-maker. There is a clear *legal division* between executive, legislative and judicial powers.

Policy presenter versus decision-maker

Article 294 distinguishes quite clearly that it is the Commission that has the formal agenda-setting power in holding the right to initiate and formulate a policy proposal (Article 294(2) TFEU) and the EP and the Council that legislate. However, with the Lisbon Treaty an OLP dossier can also be launched by an initiative from a group of the member states or the EP, on a recommendation from the European Central Bank or at the request of the Court of Justice or the European Investment Bank (cf. Article 289(4) TFEU). However, even if the Commission is still by far the primary agenda-setter, this does not mean that all proposals originate from the Commission. According to its own estimations referred to in Lord (2004: 156), only 10 per cent of first pillar acts stem from the

Commission itself.[4] The rest originate from various sources: duty to honour international commitments, improvements to keep up with technological and social change, the European Council, the Council of Ministers, the member states, the EP, the ECJ, and to fulfil Treaty obligations and/or follow up existing legislation (Lord, 2004: 156, Nugent, 2001: 236–7, 240–1). This means that the Commission is not initiating proposals in splendid isolation, but listens and is responsive to the views of the other EU institutions, particularly the European Council, the Council and the EP. In contrast to the European Council, the Council and the EP have a treaty-recognised right to request the initiation of legislation (cf. Article 241 and Article 225 TFEU respectively). In addition, the EP also uses 'own-initiative' reports to address the Commission in order to initiate new legislation (Corbett *et al.*, 2007: 239). This mechanism has been used for this purpose more frequently the last few years (Hagemann, 2009: 15). However, neither the EP nor the Council 'have the formal power to specify the contents of the proposals or to lay down a timetable for their submission' (Nugent, 2001: 239).[5]

Hence, the responsiveness towards other institutions and actors notwithstanding, the Commission possesses an important power in having the right to decide when a proposal should be launched and what the initial draft should look like, that is, 'the Commission has agenda-setting powers to structure – and restrict – the choices available to others' (Lord, 2004: 157).[6] What is more, the EP and the Council can only act on legislative issues if there is a formal Commission proposal. In sum, it is only the Commission that can present policy proposals, but what about its powers during the procedure?

Formally, the Commission has four important powers. The first is the monopoly to table policy proposals as mentioned above. The second and third are the broker and gate-keeping roles and the fourth is the right to withdraw proposals. Whereas the Commission has a strong position in the preparatory phase, its position changes once a dossier is officially launched; then the Council and the EP become the major institutional actors. However, the Commission can exert quite a lot of influence by being a facilitator and broker between the EP and the Council. Throughout the procedure, the Commission – as the only institution – attends all the central meetings in the EP and the Council, as well as the trialogues and the Conciliation Committee. Here, it participates in discussions and presents revised solutions and modified proposals (Nugent, 2001: 257, 261).[7] In fact, the Commission's role can be very important for reaching agreement between the EP and the Council (Farrell and Héritier, 2003b: 7; see also Lord, 2004: 133).

In addition, the Commission has a gate-keeping power in the sense that if it does not accept the EP's amendments, unanimity instead of qualified majority is needed in the Council. This is perhaps more accurately described as a power of delay since the EP and the Council can adopt the EP's proposal in the Conciliation Committee if the Council can agree unanimously on the EP's second reading amendments and thus sideline the Commission's rejection (see Høyland, 2006: 32). Chalmers *et al.* (2006: 153) even argue that:

there is an 88 per cent probability that a Parliament amendment will be rejected by the Council if the Commission rejects it, whilst there is an 83 per cent probability that it will be accepted if the Commission approves it.

The Commission is also free to withdraw a proposal well into the legislative process. There is, however, disagreement with regard to when the Commission's right to withdrawal expires (EP, 2007b: 7):

> While the Commission maintains its right to withdraw a proposal at any stage, the Parliament and the Council ..., consider that as soon as the Council has adopted a common position the latter – and not the commission proposal anymore – forms the basis for the further procedure. Consequently, the Commission cannot withdraw a text, of which it has not the 'ownership' anymore.[8]

It is, however, at least the case that the Commission's position weakens towards the end of the procedure and that once dossiers reach the Conciliation Committee it has no longer the right to withdraw a proposal.[9]

Altogether then, the Commission's influence is significant, but despite the above powers, it is the EP and the Council that together decide the fate of OLP dossiers as only these two institutions can decide to adopt a dossier or to veto it.[10] On this basis, it can be argued that the provisions regulating the OLP respect the separation between policy-initiator and decision-taker.

Legal distinction between executive, legislative and judicial powers?

On the basis of the above, can it also be argued that the OLP provisions are compatible with a clear legal division between the executive, legislative and judicial powers? We have already concluded that the OLP provisions ensure separation between the executive and the legislative powers. What about the last chain in the separation of powers formula? According to Article 261 (TFEU), the Treaties bestow upon the ECJ an 'unlimited jurisdiction with regard to penalties provided for in such regulations [i.e. including OLP acts]'. Hence the ECJ has the right to review acts adopted under the OLP and its decisions are legally binding.[11] Formally, there is a clear legal division between the executive, legislative and judicial branches under co-decision-making. But is this more of a checks and balances system rather than a separation of powers in the democratic meaning of the word?

> The principle of legality and the separation of powers ensure the democratic programming of administration – and not programming by experts or special interests – and bar against executive dominance and the self-acquisition of power by bureaucratic bodies. In a democratic perspective, governance is conditional on government as it is only the 'shadow' of law that governance bodies can, on the basis of delegation or audit democracy, legitimately operate.
>
> (Eriksen, 2009: 174)

Whereas there are other policy areas in the EU where the problem of executive dominance is much more pressing and where the principle of separation of powers is clearly violated, such as in the foreign, security and defence policy (see Stie, 2010b), the above citation points to the difference between a democratic reading of the principle of separation of powers and the less demanding notion of a checks and balances system, where influential actors balance each other's powers thus ensuring that none of them become strong enough to completely control the others. Although there are many aspects of this problematic,[12] I shall limit myself here to the recurring point that has a more direct impact on the OLP, namely the heavy involvement of the unelected officials/experts.

We have already discussed the lack of respect for the normative hierarchy between representatives of affected and competent parties under the criterion of inclusion. Another aspect of this debate is the fact that many of the national experts sitting in the Commission expert committees, also appear in the Council working groups as well as in the comitology committees (cf. Rhinard, 2002: 196; Egeberg *et al.*, 2006). Knowing how influential competent parties are in the overall adoption of OLP dossiers, this practice is questionable with regard to respecting the principle of separation of powers, as they act both as executives and legislators at the same time. This 'recycling system' of national experts in the lower echelons of the Commission and the Council helps to blur the relationship between the executive and the legislative branch, to the detriment of the democratic component inherent in the principle of separation of powers. What is more, the extent to which both the preparatory and the formal OLP processes are largely dominated by unelected officials (national experts as well as Council and Commission officials) only exacerbates the problem. The accountability that could potentially have been gained by the visibility of a responsible Commissioner and/or minister is in this way also not taken advantage of (cf. Cini, 2006: 16). In other words, when it is hard to find out who is doing what, it is also hard to hold decision-makers to account. This is, however, not the main democratic problem.

The main democratic problem lies in the failure of the system to (after the preparatory phase when it is common that unelected officials contribute to formulation of a dossier) sufficiently subject an OLP dossier to a process of democratic scrutiny and vindication at the level of elected representatives before the implementation phase is reached (where it is also common for unelected officials to have a significant role). More precisely, when the legislative process is not properly advanced to the political level of the Council ministers, but is primarily kept at the non-elected working group and Coreper levels, this (at least) raises the question of the extent to which the conditions for democratic control are pulverised because the public has no access to these forums. The democratic problem is, consequently, that the work of unelected officials goes largely unchecked. Whereas it can be argued that there is to some extent *political* control with experts/officials by the Commissioners and the ministers, this is not coextensive with *democratic* control. As argued above, democratic accountability is exercised in openness where the lines of responsibility can be easily grasped and identified by the public. The organisation of executive and legislative tasks is (in this manner) not in accordance with a democratic reading of the principle of separation of powers, even if it may be argued that the OLP provisions comply with a formal legal distinction.

Judicial review

> Indicator: affected Union citizens and other legal subjects can appeal to get their complaint reviewed on which basis the Court can scrutinise (and possibly sanction) not only adherence to procedural rules, but also the content of co-decision-making.

The first ex post accountability mechanism considers the possibility for a judicial review of acts resulting from the OLP. To determine whether the judiciary has the right to scrutinise OLP acts, involves considering on whose behalf this can be done. More precisely, the first aspect to look at covers the issue of *who* (or what persons/actors) can appeal. The second aspect to consider relates to *how* (or through what kind of procedure) judicial scrutiny can be conducted. The third and final aspect addresses *when* (or in what kind of cases/situations) appeals can be filed.

Article 263 (TFEU) protects the right to file an action for the annulment of directives, regulations and decisions,[13] that is, the type of acts resulting from the OLP:

> The Court of Justice of the European Union shall review the legality of legislative acts, of acts of the Council, of the Commission and of the European Central Bank, other than recommendations and opinions, and of acts of the European Parliament and of the European Council intended to produce legal effects vis-à-vis third parties. It shall also review the legality of acts of bodies, offices or agencies of the Union intended to produce legal effects vis-à-vis third parties.
>
> It shall for this purpose have jurisdiction in actions brought by a Member State, the European Parliament, the Council or the Commission on grounds of lack of competence, infringement of an essential procedural requirement, infringement of the Treaties or of any rule of law relating to their application, or misuse of powers.
>
> The Court shall have jurisdiction under the same conditions in actions brought by the Court of Auditors, by the European Central Bank and by the Committee of the Regions for the purpose of protecting their prerogatives.
>
> Any natural or legal person may, under the conditions laid down in the first and second paragraphs, institute proceedings against an act addressed to that person or which is of direct and individual concern to them, and against a regulatory act which is of direct concern to them and does not entail implementing measures.
>
> Acts setting up bodies, offices and agencies of the Union may lay down specific conditions and arrangements concerning actions brought by natural or legal persons against acts of these bodies, offices or agencies intended to produce legal effects in relation to them.
>
> The proceedings provided for in this Article shall be instituted within two months of the publication of the measure, or of its notification to the plaintiff, or, in the absence thereof, of the day on which it came to the knowledge of the latter, as the case may be.

In other words, should the Court find that there are grounds for annulment, it can rule on the basis of Article 263 (TFEU) and declare the decision void.

First, legal acts resulting from decision-making within the OLP framework are – as the Treaty article indicates – under the jurisdiction of the European Court of Justice (or the Court of First Instance). With regard to who can appeal:

> *private parties* seeking judicial review of behaviour of the EU institutions must bring their action before the Court of First Instance. However, if a *member state* or *EU institution* is seeking judicial review of such behaviour, it brings the action before the Court of Justice.
>
> (Chalmers *et al.*, 2006: 122, my emphases)

In other words, both natural[14] and legal persons can seek judicial review, that is, both affected and competent parties can file a complaint but, as we shall see, the complaints of EU institutions stand a better chance of being reviewed than the ones filed by individuals.

Second, judicial review of OLP acts can enter the legal system from two different channels. The first is the supranational channel, where a complaint is filed directly to the ECJ (or CFI) and the other follows a national route and enters the European level indirectly through Article 267 (TFEU) and the preliminary reference procedure, where plaintiffs first appeal to a national court which then refers the matter to the ECJ for a preliminary ruling.[15] In fact, the majority of Court rulings stem from the latter (Hix, 2005: 120, 128–31).[16]

Third, reasons for review are listed in the second paragraph of Article 263 and these constitute the situations and cases when appeals will be considered. These are: lack of competence, infringement of an essential procedural requirement, infringement of the Treaty or of any rule of law relating to its application, and misuse of powers. According to Chalmers *et al.* (2006: 436), there are five general principles that constitute infringement of the above grounds: fundamental rights, proportionality, legal certainty, non-discrimination and transparency. Consequently, the ECJ has the right not only to consider procedural-, but also content-specific complaints.

It should, however, be noted that it can be hard for ordinary citizens to actually get a case accepted for review with the Court. According to Chalmers *et al.* (2006: 416), the application of Article 263 first and foremost favours the member states, the Commission, the EP and the Council. These institutions are 'entitled' to bring cases before the Court. After them, the European Court of Auditors and the European Central Bank are what they call 'semi-privileged' applicants 'who are entitled to bring action but only where they do so "for the purpose of protecting their prerogatives"' (ibid; see also Costa, 2003: 744). Finally there is 'everyone else: ordinary people and corporations may bring actions under Article 230 [now Article 263] only where they have a "*direct and individual concern*" in the matter', where the Treaty formulation 'direct and individual concern' has been interpreted restrictively thus making it hard for ordinary citizens to get their case reviewed by the Court (ibid, my emphasis).[17] Moreover, it is obviously more difficult and costly for individual citizens to establish that they have been directly and individually affected by a Community legal act, than it is for firms and corporations (Costa, 2003). This is also underlined by the fact that judicial review is usually a time-consuming process and plaintiffs must be patient, 'the average length of proceedings being 21 months for direct actions and 18 months for references for preliminary rulings' (Hix, 2005: 119).

The Court's interpretation is, however, contested:

> The problem insofar as there is one, is not with the words 'direct and individual concern', but with the restrictive interpretation that the Court of Justice has given to those words. Changing that is not something that needs to be left to those with authority to rewrite the Treaties, but is the Court's own responsibility.
>
> (Chalmers *et al.*, 2006: 433)

From a democratic point of view, it is, of course, problematic that it is so much easier for institutional actors to get their case tried before the Court than it is for individuals. Chalmers *et al.* (2006: 433) do, however, note that the Constitutional Treaty would have improved the situation for individuals as Article III-365(4) (which is now Article 263 in TFEU) stated that:

> Any natural or legal person may ... institute proceedings against an act addressed to that person or which is of direct and individual concern to him or her, and against a regulatory act which is of direct concern to him or her and does not entail implementing measures.

Since this formulation in the Constitutional Treaty Article has survived in the wording of Article 263 of the Lisbon Treaty, it could become easier for individuals filing a case for annulment to have it reviewed and thus (to some extent) it may level out the playing field between institutions and individuals over time.

All in all, both natural and legal persons have the formal right to get their cases tried before the Courts even if there are some difficulties concerning the prioritisation of institutions over Union citizens as well as the fact that the proceedings are time-consuming and can be expensive, thus reducing the de facto quality of the right to judicial review. However, the possibility of judicial review is real and can thus be considered to function as a formal accountability mechanism between affected parties and the institutions, which can help to discipline the way they behave and decide.

The European Ombudsman: 'between parliamentary control and the rule of law'[18]

Indicator: affected Union citizens or other legal subjects can appeal to get their complaint reviewed on which basis the *Ombudsman* has a mandate to scrutinise (and possibly 'sanction') adherence to procedural rules and can also consider the content of OLP outcomes.

In contrast to the judiciary, the ombudsman is a soft law mechanism. The European Ombudsman (EO) was first introduced in the Maastricht Treaty and is appointed by the EP (following the Parliament's electoral cycle).[19] His task is to work against maladministration in the Union institutions and to be a mediator between citizens and EU authorities. As in the case of judicial review, the strength of the EO as an ex post accountability mechanism will be assessed with regard to the same three-fold aspects of *who* (which persons/actors) can appeal, *when* (in what kind of cases/situations) and *how* (under what kind of procedure) appeals can be made.

The legal basis designating the powers and tasks of the Ombudsman are defined in Article 228 (TFEU)[20]:

1 A European Ombudsman, elected by the European Parliament, shall be empowered to receive complaints from any citizen of the Union or any natural or legal person residing or having its registered office in a Member State concerning instances of maladministration in the activities of the Union institutions, bodies, offices or agencies, with the exception of the Court of Justice of the European Union acting in its judicial role. He or she shall examine such complaints and report on them.

   In accordance with his duties, the Ombudsman shall conduct inquiries for which he finds grounds, either on his own initiative or on the basis of complaints submitted to him direct or through a Member of the European Parliament, except where the alleged facts are or have been the subject of legal proceedings. Where the Ombudsman establishes an instance of maladministration, he shall refer the matter to the institution, body, office or agency concerned, which shall have a period of three months in which to inform him of its views. The Ombudsman shall then forward a report to the European Parliament and the institution, body, office or agency concerned. The person lodging the complaint shall be informed of the outcome of such inquiries.

   The Ombudsman shall submit an annual report to the European Parliament on the outcome of his inquiries.

2 The Ombudsman shall be elected after each election of the European Parliament for the duration of its term of office. The Ombudsman shall be eligible for reappointment.

   The Ombudsman may be dismissed by the Court of Justice at the request of the European Parliament if he no longer fulfils the conditions required for the performance of his duties or if he is guilty of serious misconduct.

3 The Ombudsman shall be completely independent in the performance of his duties. In the performance of those duties he shall neither seek nor take instructions from any Government, institution, body, office or entity. The Ombudsman may not, during his term of office, engage in any other occupation, whether gainful or not.

4 The European Parliament acting by means of regulations on its own initiative in accordance with a special legislative procedure shall, after seeking an opinion from the Commission and with the consent of the Council, lay down the regulations and general conditions governing the performance of the Ombudsman's duties.

First, and as we can see, the right to lodge a complaint with the Ombudsman is not exclusive for Union citizens, but also applies for any other 'natural or legal person residing or having its registered office in a member state'. In addition, businesses, associations and others who are registered in one of the EU's member states can file a complaint. For Union citizens, however, the right is further protected in Article 24 (TFEU). This right is also recognised in the Charter of Fundamental Rights (Article 43). In other words, the Ombudsman institution is accessible both for competent and affected parties.

Second, when receiving a complaint, the Ombudsman conducts an investigation of whether the matter is eligible for consideration as a case of *maladministration*. His main task, then, is to deal with instances of 'maladministration' – a word that is not further specified in the Treaties and which consequently has been up to the interpretation of the Ombudsman institution itself to define (Chalmers *et al.*, 2006: 337). Over the years, the Ombudsman has defined maladministration as administrative irregularities, unfairness, discrimination, abuse of power, failure to reply, refusal of information and unnecessary delay.[21] With regard to the OLP, it is mainly instances of access to information/documents that are relevant. One of the main goals the first Ombudsman took upon himself was to establish a written Code of good administrative conduct;[22] this was achieved in 2001. The introduction of the Code implied a departure from the rather legalistic definition of maladministration in the initial years (1995–1997) of the Ombudsman's office, to cover non-legal aspects as well (Chalmers *et al.*, 2006: 339–41).

Third, and in contrast to the Courts, to file a complaint with the Ombudsman is a much simpler and cheaper affair. Not only is the service free of charge, it is also fast and flexible in comparison with a court procedure which can drag out. The complaint form can be accessed and filled out electronically on the Ombudsman's website, or be sent via mail or fax in all of the official EU languages.[23] It should be noted, however, that the majority of the complaints received by the Ombudsman fall beyond his 'jurisdiction' and are therefore rejected. The Ombudsman can only investigate and assess the decisions and activities of the EU institutions (with the exception of the ECJ and CFI when acting in their judicial role), including decision-making processes that originate within the OLP.[24] In addition, he cannot investigate instances that are already dealt with by the Court.

However, once a complaint falls within his 'jurisdiction', the EO informs the given institution about the matter. If the institution does not respond to or resolve the issue, the Ombudsman tries to find a 'friendly solution'. If this is successful, the case is closed by a 'reasoned decision'. If, however, a 'friendly solution' is not successful, he can either make a 'reasoned decision' entailing a 'critical remark' or proceed to make 'draft recommendations', and finally, if the institution does not accept the 'draft recommendations', he can file a special report to the EP, which is made public. This is the strongest 'weapon' or 'sanction' the Ombudsman has at his disposal (cf. Article 228, TFEU). In addition and in contrast to the Court, which only reviews cases when complaints are being filed by other actors, the Ombudsman is not constrained by such regulations and can on his own initiative start investigations of maladministration.

There are obvious limitations to the EO's 'sanctioning' mechanisms as they are not legally binding or judicially enforceable. However, this lack of legal power has also proven to be a strength, in the sense that the Ombudsman can allow himself to be more direct in his recommendations. Moreover, the route to his success lies both in the working method as well as the legitimacy of his position and the validity and general endorsement of the principles he is defending:

> On the one hand, acting like a Court, he interpreted the cases submitted to him in a teleological way, in order to build a demanding doctrine of 'good administration'. On the other hand, acting as a parliamentary organ, and with the strong support of the EP, he used his powers of inquiry and proposition to suggest wide-ranging reforms of European governance.
>
> (Magnette, 2003: 690)

Consequently, even if limited, the EO's powers:

> give him the opportunity to combine the instruments of parliamentary scrutiny and judicial control in an original way. Moreover, given the hybridity of its status and role, the Ombudsman is well equipped to scrutinize those agents that cannot be submitted to classic parliamentary controls without losing their independence, and thereby help to reconcile delegation with parliamentary democracy.
>
> (ibid: 678)

His success is, of course, conditional and dependent upon balancing good relations with the institutions, on one hand, and maintaining a certain independent legitimacy that gives room for manoeuvring, on the other. To publicly announce the results of his investigations can be quite a successful strategy, as the institutions are (presumably) to some extent concerned about their reputation and want to avoid public shaming. The Ombudsman has in this way established a practice where EU institutions are forced to justify their positions and, if not complying with the EO's recommendations, must provide explanations for why they do not find them convincing.

As mentioned, transparency has been one of the Ombudsman's main concerns and has contributed considerably to easing the accessibility and quantity of documents available to the public. The first Ombudsman, Jacob Söderman, made this issue one of his main tasks and worked to improve the standard of openness in the Union: 'The EO is playing an important role in correcting negative rooted institutional practices in the field of access to documents, and its decisions show that he is very much inclined towards openness' (De Leeuw, 2003: 347; see also Magnette, 2003: 689; Curtin, 2007b). Moreover, Magnette (2003) argues that the Ombudsman is an important accountability mechanism, particularly in very complex and technical polities such as the EU. The Ombudsman also contributes as a complementing control mechanism along parliamentary and judicial control because the EP does not have the same powers as those held by national parliaments.

In this regard, an example of how the EO works can be illustrated by how he (as well as the EP) has pressured to open up Council deliberations. In 2003,[25] prior to the Council transparency improvements in 2006, the Ombudsman received a complaint from an MEP and a youth politician arguing that the Council's rules of procedure were not in line with Treaty obligations to 'act as openly as possible and as closely to the citizens as possible' as they supported closed rather than open decision-making meetings. More specifically, the complainants argued that the Council is an institution that takes decisions directly affecting the lives of Union citizens and thus has an obligation to meet and discuss legislative proposals in public. They referred to the (then) draft Constitutional Treaty, which confirmed their interpretation that all legislative processes should be open. The Council, however, argued that the Constitutional Treaty had not yet been ratified; that the phrasing in the Treaties about openness and confidentiality of decision-making was more of a programmatic principle (to be realised some time in the future) than an exact rule; and consequently that its rules of procedure were neither violating the Treaties nor did they constitute a case of maladministration.[26] Moreover, the Council further responded that 'the way in which the Council organised its internal procedures and the fact that the public was not admitted to all Council meetings was one of the political choices made by the Council when it organised its internal procedures' (EO, 2005b). In other words, the Treaty principle of open decision-making had nothing to do with the way the Council organised itself. The Ombudsman supported the complainants in his draft recommendation and upheld his position in a Special Report to the EP, which adopted it unanimously. The Ombudsman's recommendation has since been followed up by an EP committee report urging the Council to open up its proceedings, especially when acting as a legislature (as it does under the OLP). The point is that the burden of evidence is reversed and it now falls on the Council instead of the complainants to show that they are in line with Treaty obligations.

We can, of course, only speculate about what kind of impact (if any) the Ombudsman played in the Council's decision to revise its rules of procedure in 2006,[27] which committed the Council to more open deliberations, but it is not unlikely that the Council also felt the pressure from a thoroughly written EO report that is publicly accessible. From a deliberative democratic perspective, however, it illustrates how the Ombudsman offers affected (and competent) parties an institutionalised procedure where the governing institutions are 'forced' to provide reasons and justifications for their decisions. In this sense the EO represents an important institutional mechanism for upholding the practice of account-giving and publicity.

Conclusion

In this chapter I have looked at four mechanisms that help to neutralise asymmetrical power relations. The first two are ex ante mechanisms, which consider the potential intelligibility of the OLP and the degree to which it is possible to argue that it respects the principle of separation of powers respectively.

With regard to the first, it could be confirmed that the OLP is, overall, intelligible, but not without critical weaknesses. Article 294 (TFEU) prescribes a sequential organisation of the procedure with the possibility of up to three readings and where the formal processes take place in the EP and the Council respectively as well as in the Conciliation Committee if necessary. The two Joint Declarations (and to some degree the institutional actors' rules of procedure), on the other hand, describe a procedure where the informal trialogues play a highly significant role and which consequently challenge the formal processes in the EP and the Council. Moreover, the character of the OLP is now more correctly described as simultaneous not sequential. Hence the intelligibility of the OLP is seriously weakened, but when seen in isolation it is still possible to say that the indicator is complied with. However, when seen in relation to the other indicators and criteria, this conclusion may change. I come back to this in the next as well as in the concluding chapter.

With regard to separation of powers, the picture is mixed. Whereas the Commission is the only institutional actor with the power to table policy proposals and which also wields considerable powers during the formal OLP process, it is nevertheless the EP and the Council that have final decision-making power. In this sense, the policy-presenter is not the same as the decision-makers. From a formal perspective, there is also a clear legal distinction between executive, legislative and judicial powers under the OLP. However, when we look at the heavy influence of unelected experts and officials throughout the procedure – from the preparatory/agenda-setting phase to the official launch of OLP dossiers and the main legislative process – the separation of powers is blurred as many of the same experts participate both in Commission expert groups, Council working groups and comitology groups. But even more importantly, the internal organisation of the Council (where the working groups and Coreper are dominating the preparation and the formation of the Council position and where the meetings at the ministerial level are not conducive to a democratic testing of proposals) consequently violates the principle of separation of powers as the lines of responsibility become unclear.

The latter part of this chapter assessed the conditions for judicial review and the possibility for an Ombudsman inquiry when affected parties see their rights violated by the adoption and implementation of OLP dossiers. These two ex post accountability mechanisms both prove to be met satisfactorily even if appeals for judicial review are accepted more readily for EU institutions than for Union citizens. This inequality is exacerbated by the fact that judicial review is a protracted process and consequently easier for the institutions to cope with than for individual persons. In this case, the Ombudsman can be seen as an alternative as the proceedings are easier, cheaper and less time-consuming. The EO is only a soft-power mechanism and does not have the same power as the Court, but its draft recommendations and reports can be quite effective in publicly highlighting and shaming instances of institutional maladministration. The EO has, in particular, focused on the issue of transparency and has presumably helped to open up EU institutions. Moreover, in exercising their powers both the Court and the Ombudsman provide settings and procedures in which decision-makers can be held accountable and where they have to justify their position in an open manner. In this sense they contribute to fulfil the deliberative democratic emphasis on institutionalising and maintaining publicly accessible practices of justification and accountability.

9 Decision-making capacity

Introduction

Under the fifth and final criterion, the task is to evaluate the OLP's decision-making capacity. By this is meant that decisions are also actually discussed and taken within the basic parameters of the OLP framework as set up in the Treaties. This is important, as the strength of the conclusions reached about the democratic qualities of co-decision-making under the other criteria is also dependent on whether it can be argued that decision-making actually takes place within the assigned framework of Article 294. If not, then decisions taken in other forums and/or by other actors are simply dressed up and rubber-stamped as OLP acts.

On the other hand, this does not mean that all types of informal meetings are automatically deemed illegitimate. What is more, it is also acknowledged that the formal institutions prescribed and described in the legal provisions are not, in most cases, ready-made recipes which do not require practical interpretation when applied to actual empirical circumstances (cf. Farrell and Héritier, 2004). This is also the case with regard to co-decision-making, and Shackleton (2000: 333; cf. also Farrell and Héritier, 2004) remarked that the inception of the procedure only offered 'a framework which had to be filled in through practice'. However, the concern here is how the type, scope and scale of such formally unauthorised meetings can potentially undermine the formal processes by representing the locus where deals are actually struck.

The point is that the actual practice of the OLP should be in line with the *intention* of the formal legal provisions and that informal structures have not replaced (and thus made redundant) the formal set-up. As far as I can see, there are two 'candidates' that may represent a competing decision-making alternative. The first 'candidate' is the Council, which is by far the more powerful institution in the Union and, as we have seen in Chapter 4, did not originally interpret Article 294 as a genuine co-legislative decision-making framework (Corbett *et al.*, 2007: 225). Despite the fact that the Lisbon Treaty doubled the number of policy areas assigned under the OLP and thus secured the EP more power, it is not unlikely that the Council is in fact the stronger party in co-decision-making, as it is in all other areas of EU policy-making. The second potential 'candidate' is the trialogue arrangement. Throughout the book, the importance of trialogue meetings to the functioning of co-decision-making has been emphasised. The question is therefore whether they have become so influential that it is reasonable to argue that they now dominate (or even dictate) the direction and content of the outcomes of the formal processes in the EP, the Council and the Commission.

> Indicator: there is *absence of informal networks* where powerful actors 'pre-cook' proposals that are not subsequently sanctioned by democratic organs and which consequently have taken over as the locus where actual co-decision-making takes place.

Is the Council a competing decision-making arrangement?

With the introduction of the OLP, a major question was obviously whether the Council would commit to the *co*-legislative scheme stipulated in the letter of Article 294 (previously Article 189b TEC) or whether it would try to by-pass the EP (Farrell and Héritier, 2004). Historically, the relative power balance between the two legislatures was (and still is), to say the least, uneven. Up until the Single European Act (which introduced the cooperation procedure) and particularly after Maastricht and the inception of the OLP, the Council's 30–40-year legislative monopoly was finally challenged (cf. Chapter 4). The development of a closer relationship and an atmosphere of cooperation between the EP and the Council were consequently not self-evident from the beginning and the first few years were also characterised by less cooperation and more antagonism between the two legislatures (Shackleton, 2000: 236).

As we have seen, under the Maastricht provisions of 'co-decision I' the formal cooperative incentives were also fewer than under the current 'co-decision II' regime.[1] At this point, the Council was allowed to re-introduce its Common Position if agreement on a Joint Text could not be achieved in the Conciliation Committee. The only way the EP could reject the reintroduced Common Position was if it managed to drum up an absolute majority. This possibility could be seen as an incentive for the Council to decide to make less of an effort during the conciliation process with the EP or, in the words of Warleigh (2003: 31), the risk was that the 'Common Position ... [would] always be a bargaining position rather than a legislative text.' The Council only used this right once in the Voice Telephony Directive in 1994, but the EP managed to gather an absolute majority and turned the Council's Common Position down (Shackleton and Raunio, 2003: 171; see also Burns, 2006: 235). Through this action the Parliament sent an important message to the Council that it could and would block legislation if the Council did not work harder to find agreement with the EP during conciliation (Shackleton, 2000: 237). Or in the words of Burns (2006: 236) 'the EP showed that when confronted with a "take-it-or-leave-it offer" from the council under co-decision I it would automatically reject legislation.' The 'warning' paid off and throughout the 'co-decision I' period only two dossiers failed.[2] By the time the negotiations on the Amsterdam Treaty started, the Council had anticipated and acted on the assumption that the right to re-introduce the Common Position would be repealed (ibid).[3]

The current 'co-decision II' provisions are consequently more in line with the co-legislative arrangement underpinning the formal letter of Article 294 in the sense that they do not make the Council await the possibility of side-stepping the EP after a 'failed' conciliation, but rather ensure that conciliation efforts are not rendered redundant (Warleigh, 2003: 31–2; Häge and Kaeding, 2007: 344). Moreover, the right to veto a position:

> gave Parliament a bargaining position which it hitherto lacked regarding Community legislation, and was of fundamental importance to perceptions of its role – it could no longer be accused of lacking teeth. It is not that the Parliament necessarily says no (in fact, it has only done so on six occasions between 1994 and 2006) but rather that the Council recognises that this is a possibility if it does not negotiate seriously. The very fact that there have only been six rejections is a testimony to the fact the Council has been willing to look for common ground with the Parliament and to modify its common position in the process.
>
> (Corbett *et al.*, 2007: 216)

The development and expansion of the trialogue system is in itself an indication of the Council's willingness to respect the EP's position under the OLP. This commitment is also reflected in the two aforementioned Joint Declarations on the workings of the trialogue meetings (see below). The fact that the EP is not only a rubber-stamp body signing whatever the Council decides is also (even if implicitly) corroborated by the many game-theoretical studies that I mentioned in Chapter 1, whose objective is to measure how the OLP has affected the inter-institutional balance in the EU. In this inter-institutional 'game', the EP comes out favourably because it is (usually) less impatient than the Council due to the latter's cycle of six-month rotating presidencies (see e.g. Häge and Kaeding, 2007). Consequently, if the EP did not matter at all, why would these researchers bother setting up fancy equations and continue to discuss these issues year after year? Hence, the fear that the Council would be the actual decision-maker with the EP simply approving the former's decisions is not correct.

Do the trialogues represent a competing decision-making arrangement?

The problem of cooperation between the Council and the EP no longer exists. However beneficial the cooperative atmosphere between the EP and the Council has now become, the very institution of trialogues needs to be addressed with regard to whether it challenges the formal OLP arrangement and thus represents a competing decision-making structure. As shown throughout the book, the trialogues have a central position in co-decision-making and many see them as a precondition for the successful completion of decision-making, especially after first reading (e.g. Farrell and Héritier, 2003b).[4] Having said this, it is obviously not easy to determine exactly when an informal practice constitutes a competing decision-making network or arrangement and when it is not, especially since the trialogues are, as we have seen, now more or less semi-formalised in the two Joint Declarations.

If we take a look at how Article 294 describes the procedure, it is (as we have seen in the previous chapter) *sequential* in character. The EP and the Council (with the Commission) are supposed to respond to each other's suggestions in a *relay* fashion and they can afford to do so in up to three readings if disagreements cannot be settled in the first. Quite a lot of space is devoted, in Article 294, to describing how and when each reading can be terminated and when a proposal either fails or is adopted, that is, what kind of majorities are needed for an act to pass in each house as a collective body, *not* by the discretion of only a handful of people. Hence the intention behind Article 294 seems to be quite clear. The trialogue system obviously alters this framework, as co-decision is no longer correctly described as sequential, but rather *simultaneous* in character. In this sense, then, the character of the procedure is no longer in line with the provisions of Article 294. Does this situation turn trialogues into a competing framework where the trialogue participants are the actual decision-makers thus rendering formal treatment in the EP and the Council a charade?

Conciliation trialogues

Originally, trialogues were only conducted prior to meetings in the Conciliation Committee. In the Treaties, the set-up of the conciliation process is (still) confined only to the formal meetings in the Conciliation Committee (cf. Article 294 (10–12) TFEU). Consequently, and as a result of many fruitless discussions in these meetings (which can contain over 100 people in the room) trialogues were introduced in order to make conciliation work more smoothly and to fill the void in the Treaties on what should happen in the time span between a Council rejection of the EP's position in the second reading and the scheduling of a Conciliation Committee meeting.[5] Trialogues were thus established to prepare and anticipate the results of Conciliation Committee meetings proper. Many contend, however, that the trialogues dominate the conciliation process and that the Conciliation Committee meetings have become a mere formality where decisions already taken in trialogues are simply being 'let through on the nod'.[6]

As mentioned, it is commonly argued that the trialogues are necessary in order to reach agreement on OLP dossiers, i.e. they function as settings where the institutions can sound out and discuss differences, flexibilities and limits in a more frank atmosphere (Farrell and Héritier, 2004: 1198). Hence the purpose is to prepare decisions, not to take them. However, the working method applied in *conciliation* trialogues appears to be adopted from the Council's practice, namely the 'A' and 'B' point procedure.[7] This supports the above claim as it indicates (as in the Council) that during the conciliation process, many decisions (the 'A' points) are not only prepared, but are actually also taken in trialogues and thus only rubber stamped in the formal Conciliation Committee meetings. The following quotation gives an indication of the relationship between conciliation trialogues and the Conciliation Committee as well as an indication of the relationship between those who participate in conciliation and those who do not:

> although it is normal for trialogues and other informal contacts to narrow the ground before the Conciliation Committee meets (and sometimes to resolve all outstanding points so that conciliation becomes a mere formality), decisions sometimes have to be made on the spot in the Conciliation Committee, without the chance to refer to capitals.
>
> (Bostock, 2002: 221–2)

The same is indicated by Farrell and Héritier (2003b: 21–2): 'It is fair to say that the Conciliation Committee meetings are increasingly *pro forma;* much of the real politics and bargaining takes place in the informal trialogues that precede them.' Moreover, according to Shackleton and Raunio (2003: 177), the conciliation trialogue:

> is a meeting where ideas can be exchanged without formal decisions being taken, each side accepting to refer back possible solutions to their delegation. Should the discussions be sufficiently successful, the full conciliation committee can be avoided and the issue concluded in writing or as an 'A' point, without discussion, at a conciliation on another issue.

In sum, seen from the perspective of 'A' points, it seems highly unlikely that the Conciliation Committee is the main decision-making locus.

So what about 'B' points, are they still treated more substantially and not just voted on in the Conciliation Committee? In its Codecision and Conciliation guide, the EP (2009a: 22) states that:

> Normally, a short trilogue meeting is held just before the meeting of the Conciliation Committee and sometimes the Conciliation Committee meeting itself is interrupted for negotiations in trilogue to clarify the situation, to find mutually acceptable compromises, and to avoid misunderstandings between the delegations.

From this quote one can at least get the impression that 'B' points are *also* (and to a substantial degree) dealt with in the trialogues that take place between the formal Conciliation Committee meetings and that the latter seem more like an arena where what has been agreed upon in trialogues is reduced to formal approval and the casting of votes in the Conciliation Committee.

Having said this, it should also be underlined that trialogue participants are more constrained by formal rules during conciliation than in the preceding stages. First, during the conciliation process both the EP and the Council appoint delegations to ensure and monitor that the behaviour of trialogue participants is in line with the respective 'mother' institutions (cf. also Rasmussen, 2005). Moreover, the whole purpose of the delegations is to keep an eye on the trialogue participants and provide them with a mandate on how to proceed. As we have seen in Chapter 5, the EP delegation receives information about when and where trialogue meetings are being held and who participates (EP, 2009a: 20). The time of the trialogue meetings is also published in the EP's list of daily meetings (ibid). In the Council, the conciliation delegation amounts to the permanent representatives in Coreper. In addition and more importantly, both the EP and the Council delegations must approve the results of the trialogue meetings (ibid: 20). In this sense the conciliation trialogues are better controlled than the first and second reading trialogues (Rasmussen and Shackleton, 2005), not least since the various positions are, by then, (arguably) well known and articulated.

Second, trialogue participants have also traditionally been constrained by the limitations on what kind of amendments and changes can be made to the dossier in the Conciliation Committee (i.e. that they should, ideally, not be completely new, but based on the EP's second reading amendments to the Common Position[8]). As one of the interviewees in Rasmussen and Shackleton (2005: 12) argued:

> In conciliation everybody knows what the starting point is: the amendments of the Parliament and the common position of the Council, and it's written down … we know where everybody is coming from and we know where they end up, but at first reading the ground is totally unclear and indeed the rapporteur is able to determine what the ground should be.

In other words, by the time the process reaches conciliation, the positions and discussions have (presumably) already revealed where the difficult points lie, which henceforth limits the trialogue participants' room for manoeuvring. Moreover, in this regard the EP's and the Council's respective conciliation delegations represent an institutional mechanism that ensures a much tighter monitoring of the trialogue participants during the conciliation process than in the first and second readings. In the EP, the conciliation delegation has specifically assigned formal rules dictating how the trialogue participants shall behave; they receive both written and oral mandates and they have a duty to report on the proceedings and indicate how they are to move forward (EP, 2009a: 17–21). In the Council there are seemingly no particular rules for conciliation processes, but given that Coreper is in charge, the standard operating procedure of intra-institutional co-ordination applies.

The only fly in the ointment is that the importance of the conciliation procedure is (arguably) becoming increasingly obsolete as the overwhelming majority of OLP dossiers are now concluded after the first reading and hardly any dossiers reach the conciliation stage anymore. As shown in Chapter 4, in the fifth parliamentary period (1999–2004), 28 per cent of all OLP dossiers were closed after first reading (115 dossiers), whereas in the sixth parliamentary term, 72 per cent of the dossiers were concluded at first reading (321 dossiers). And the early agreement trend only continues, in the seventh parliamentary period (from 14 July 2009 till 14 March 2012) 77 per cent (151) of the dossiers were concluded at first reading.[9] Moreover, the costs of rejecting a proposal are probably higher at the conciliation stage than in the previous readings (even if the emphasis on early agreement make the stakes high from the very beginning) as the whole legislative process would have to start all over again (Rasmussen, 2005: 1025). This is confirmed by the very small number of dossiers that have been rejected after conciliation.[10]

First and second reading trialogues

The trialogues are no longer merely a trait of the conciliation process, but take place and are strongly encouraged throughout the procedure.[11] There is also an explicit commitment to reach agreement as early as possible, preferably during the first reading.[12] In comparison to the conciliation trialogues, the first and second reading trialogues are, however, less controlled as there are no conciliation delegations to monitor and keep them accountable. The scope for independent action by trialogue participants has also proven to be wider (Rasmussen and Shackleton, 2005: 12–3), especially in the first reading where the positions are usually more malleable and less fixed as: 'both institutions are negotiating without either having finalised a point of departure for negotiation. Trialogue at this stage is even more informal than during conciliation' (Bjurulf and Elgström, 2004: 261). Moreover, during the first reading the trialogue participants can take advantage of the fact that there is not, at this point, an official EP or Council position to defend (which is the case in the second reading). In addition, in Parliament there is no restriction on the kind of amendments that can be presented and only a simple majority is needed in order to adopt a proposal (see also Rasmussen, 2007: 6).

The problem with first reading trialogues is that the asymmetrical information gap between the Council and the EP actors is wider than in the subsequent readings, as the Council is not open to MEPs (or to citizens in general) either at the level of the working parties or at the Coreper level. Hence, the MEPs who are not participating in trialogues must rely on information from their fellow MEPs who do take part, or on personal contact with Council or other involved actors. Unlike conciliation trialogues, the responsible EP committee does not have the same capacity or mandate to monitor the trialogue participants in the first and second readings as the conciliation delegation has during the conciliation phase. As discussed in Chapter 6, the EP's Code of Conduct[13] for trialogues states that key MEPs should respect their delegated authority in discussions with the Council and the Code specifically emphasises that:

> As a general rule, the amendments adopted in committee or in plenary shall form the basis for the mandate of the EP negotiating team. The committee may also determine priorities and a time-limit for negotiations. In the exceptional case of negotiations on a first reading agreement before the vote in committee, the committee shall provide guidance to the EP negotiating team.[14]
>
> (European Union, 2010: Article 267 (TFEU))

The Code also underlines that the rapporteur has a duty to keep the other committee members informed (both in writing and orally) – first and foremost through the committee chair, the group coordinators and the shadow rapporteurs. Moreover, there is an emphasis on the importance of the committee in that it 'shall consider any agreement reached or update the mandate of the negotiating team in the case that further negotiations are required' (EP, 2009b: 44). In addition, the team shall always report back regularly on the state of negotiations and distribute texts to the whole committee (ibid). Hence there is an intention or attempt to make the trialogue participants accountable to the responsible committee on a par with the role of the EP conciliation delegation.

In this regard, Rasmussen and Shackleton (2005: 19–20) note that 'the provisions are not binding but they establish a yardstick against which the behaviour of negotiators, above all, their ability to keep the confidence of their colleagues, both inside and outside the committee, can be judged'.[15] However, we know from the above that this principle of awaiting trialogue contact until the process has started in the EP committee is being violated. In many cases the rapporteur has already been in touch with the Council through trialogues and the committee is often presented with an already agreed deal that can be hard to reject. In this regard, Bunyan (2007b) draws attention to a list of OLP dossiers in justice and home affairs policy areas formerly figuring under the third pillar but, since 2005, dealt with under the OLP.[16] These controversial dossiers have been discussed and agreed before they have been properly dealt with in committee.

Even if the Council is the champion and defender of the trialogue system, early agreements are also affecting its own internal balance. Whereas the Council representatives not participating in trialogues probably have more information overall than their fellow MEPs (Benedetto, 2005: 70–1), this is not the same as actually being in the room when agreements are reached. In the Council, trialogues put the Presidency in a particularly favourable position as 'it can present its version of the Council position to the EP, and report back its version of the EP response. Its room for tactical manoeuvring is therefore substantial' (Bjurulf and Elgström, 2004: 261; see also Farrell and Héritier, 2003b and 2004). Hence, 'It may potentially use this leverage to affect other actors' perceptions of what is possible and what is not, and thus bring through outcomes which reflect its own preferences rather than the preferences of the Council as a whole.' (Farrell and Héritier, 2003b: 29). Moreover, early agreements presumably make the room for manoeuvring even wider simply because the positions have not yet been firmed up and a Common Position has not yet been formulated. However, the Presidency's independence has obvious limits and is dependent upon trust from the other national delegations in order for the latter to accept 'the intrinsic disadvantage involved in so much activity taking place outside their direct vision at meetings to which they are not invited' (Shackleton and Raunio, 2003: 175). As in the conciliation process, the trialogue participants (i.e. the Coreper representative of the country holding the Presidency) cannot (obviously) deviate too much from the positions of his/her fellow Coreper representatives.

Trialogues in general

From Chapter 6, we already know that trialogues upset the internal division of labour between the committees and the plenary sessions in the EP. Trialogues make non-attached MEPs in particular, but also MEPs from small political groups, less able to influence dossiers due to the informal liaisons between key persons in the largest political groups in the EP and the permanent representatives/Presidency in the Council.[17] Deals are struck in trialogues and presented as 'take-it-or-leave-it' offers to the other MEPs (Foilleux *et al.*, 2005: 622). This tendency also has a national dimension in the sense that the member states try to influence and form alliances with 'their' MEPs through the national delegations. Here the larger member states will presumably be more successful than the smaller member states, as size and resources are likely to matter:

> This emerging trend may have important repercussions for decision-making. It allows larger member states a 'second bite at the cherry'. Even if they find themselves marginalized in discussions over a specific piece of legislation within the Council, larger member states may be able to mobilize support among MEPs so that the piece of legislation in question is amended to their satisfaction, or rejected.
>
> (Farrell and Héritier, 2003b: 27–8)[18]

It is also a widely recognised problem that when time is tight and the general level of workload is high, as is the case for legislative actors in the EU, this creates and/or deepens information asymmetries between those who are directly involved in a dossier (like the rapporteur, the committee chairs, the political group coordinators and the Council Presidency) and those who are occupied with other cases (the rest of the MEPs and Council permanent representatives/ministers). The overwhelming majorities with which OLP dossiers are adopted in the EP can be seen to support the argument that the shortage of time (and perhaps also an interest in a particular dossier) tends to make the MEPs vote according to political group lines (cf. Chapter 6; Settembri, 2007).

What is more, the Joint Declaration also seems to suggest that the balance between formal institutions and trialogues tilts towards the latter with regard to where agreements are formed and reached. The indication of this is that the former (arguably) more often become settings where these items only await formal approval and not substantial treatment. Here is only one example:

> Where an *agreement* is reached through informal negotiations in trilogues, the chair of Coreper shall forward, in a letter to the *chair* of the relevant parliamentary committee [i.e. not the committee as such], details of the substance of the agreement, in the form of amendments to the Commission proposal. That letter shall indicate the Council's willingness to accept that outcome, subject to legal-linguistic verification, should it be *confirmed* [i.e. not discussed and then decided] by the vote in plenary.
>
> (European Parliament, Council, Commission, 2007: rule 14, my emphases)

Consequently, even if lacking *formal* Treaty recognition and *explicit* decision-making power in the Joint Declarations, it does not seem unlikely that substantive decisions are taken here. What is more, the impact of trialogues on the overall outlook of OLP dossiers is stronger if agreement is reached during first, rather than in second and third, readings. In this way the informal practices seem (at least to a considerable extent) to empty the formal processes of decision-making clout.

This is also indicated by how the question of whether trialogues should be further institutionalised and formalised is perceived in the EP. The Parliament has been, and still is, more ambivalent than the Council about the role trialogues have obtained. On the one hand, the trialogue system is a source of influence for the EP, but, on the other hand, the Parliament's legitimacy suffers due to its lack of transparency and accountability (Farrell and Héritier, 2003b: 23). This is particularly a problem with regard to first and second reading trialogues, and it has therefore been suggested that these meetings are formalised and institutionalised in line with the methods used by the conciliation delegation to control the conciliation trialogues. In their study, Rasmussen and Shackleton interviewed central OLP participants on this issue. According to one respondent, the rapporteurs 'do not wish there to be greater institutionalization at the moment, because they know this is what gives them ultimate power … it seems to me that it is a highly dubious instrument' (Rasmussen and Shackleton, 2005: 13). Moreover, another respondent contended that:

> There are several reasons why many MEPs do not like to go to conciliation because for them conciliation means a loss of power … in conciliation we have rules, we have formal delegation, regular meetings, everyone is concerned, whereas in first and second reading there is an absence of rules.
>
> (Rasmussen and Shackleton, 2005: 14).

What is more:

> There are likely to be various circumstances where MEPs are willing to make a trade-off between more open procedures and the desire to achieve a result more quickly. At the same time, Parliament has resisted the Council argument that such trilogues are the normal way of agreeing all legislation and wants Council to accept the norms of an open deliberative chamber by attending and participating in committees.
>
> (Shackleton and Raunio, 2003: 179)

In other words, the question is if we have come to the point now where the trade-off has gone in favour of efficiency given the overwhelming number of dossiers now completed at first reading. If that is the case, this is in line with the Council's wishes as it views trialogues as a:

> way to speed up legislation, and make it more efficient. Although it has recognized their existence in the Amsterdam Treaty, it is unwilling to have the Treaty prescribe any specific formal structure for them, in case that limits their flexibility.
>
> (Farrell and Héritier, 2003b: 23)

Consequently, it is one thing if trialogues are confined to the conciliation process, but knowing that they take place continuously throughout the process, that they are regarded by the participants as being essential for reaching agreement and that it is most likely here that deals are struck, at least indicates that actual decision-making also happens within them. Seen from this perspective, the above supports the conclusion that trialogues represent a competing or at least a parallel decision-making structure where the actual decisions are taken. Chalmers *et al.* (2006: 155, my emphases) have reasoned along the same lines and state even more explicit and forcefully that:

> There is, in all this a sidelining of checks and balances and a lack of formality and transparency. *A division is made between formal and substantive decision-making, with the locus of substantive decision-making being hidden away.* Whilst formal decision-making takes place in the Council or in Parliament committees, in many instances substantive decisions are vested in these informal arrangements. The formal procedures do no more than *rubber stamp* the agreements. Only very well connected actors have the opportunity to lobby these informal processes because only they can know where they are taking place or who is important within them. Furthermore, only they will have the resources to arbitrate between these centres of power, lobbying both central protagonists in the trilogue and other important actors in the Council, the Parliament and the Commission.

This view is also supported by Farrell and Héritier (2003b: 22):

> Trialogues are not formally binding; neither the Council nor Parliament is obliged to adhere to agreements reached in these meetings. However, because Council and Parliament engage with each other repeatedly in the legislative process, it is usually in their interest to make agreements stick; otherwise the defaulting party is likely to lose credibility, and to be punished in future interactions.

In sum, the above points support, rather than disconfirm, the assumption that trialogues represent a competing decision-making arrangement. The trialogue system in itself has at no point been tested and scrutinised by a democratic process, but rather introduced incrementally and through the means of inter-institutional agreements that are not always in line with Treaty prescriptions (cf. Curtin, 2007b). In this sense, it is clearly challenging and altering the intention of the formal set-up of how the OLP processes are supposed to unfold.

Conclusion

In this chapter, the focus has been on the decision-making capacity of the OLP. To assess this criterion we looked for competing decision-making frameworks. In this regard two candidates were reviewed: the Council and the trialogues. Whereas the theory that the Council does not (alone) represent a competing framework of co-decision-making can be corroborated, the conclusion on the trialogue system is less straightforward.

From the above it seems safe to argue that the Conciliation Committee is not functioning as the locus where the final outstanding points are discussed and possibly agreed. Rather, it has apparently been reduced to a mere voting arena whose (more or less) sole purpose is to formally approve the Joint Text. In contrast the trialogue has become the inter-institutional setting where the substantive discussions are conducted and where deals are struck. However, and in contrast to the first and second readings, during conciliation the trialogue proceedings have more formalised rules and institutionalised practices for mandating and controlling the respective trialogue-participants through the EP's and the Council's respective delegations. This compensates, to some degree, for the fact that the Conciliation Committee has partly been emptied of purpose and function as it is intended and prescribed in the Treaties. But again (and unlike the power of the conciliation delegation to mandate and control the trialogue-participants during the final stages of the OLP), the guidelines on first and second reading agreements are not binding upon the participants.

If we look at the trialogues merely from a power perspective, it is difficult to reach a decisive conclusion. On the one hand, the first (in particular), but also the second reading trialogues are hard to control for the non-participating Council and EP representatives. Hence from the first- and second- reading- perspective, trialogues could easily be perceived as a competing decision-making framework, particularly if we also take into consideration the extent and scope to which trialogues are now conducted. On the other hand, the conciliation trialogues are much better controlled than the first and second reading trialogues due to the respective EP and Council delegations. These delegations thus represent an accountability mechanism that presumably makes it easier for both the EP and the Council as collective legislative bodies to check and hold the trialogue participants to account. What is more, from a pure power perspective, it would also be pertinent to underline that, in the end, it is the formal settings of the EP and the Council that are the ultimate decision-making arenas, as it is only the Parliament and the Council as collective bodies that can adopt or reject proposals. In sum, from this perspective, the conclusion is unsettled.

However, the task here is to frame the question of decision-making capacity in a democratic perspective and from this vantage point, the situation is more clear-cut. The trialogues obviously by-pass the formal settings where the decisions are (even if not perfectly) discussed, challenged and taken in more inclusive and open ways. What the trialogues do – especially in the first and second readings – is to disconnect (to a large extent) the link between the deliberative opinion- and will-formation processes that take place in the EP committees and plenary sessions and the subsequent decision-making moment when the formal votes are cast. The trialogues constantly interrupt this process and effectively remove the substantial opinion- and will-formation process to a locus where the general public does not have access, either during or after the process is finished. This point does not apply to the same degree to the Council, since its internal organisation already fails to ensure a democratic opinion- and will-formation process as we have seen in Chapter 6. However, the use of trialogues only adds to the impermeability of Council political processes. In sum, from the perspective of ensuring popular control of public power, the extensive use of trialogues is detrimental to the whole idea of rule by the people.

10 OLP: democratic decision-making or technocracy in disguise?

Introduction

From its inception in the Maastricht Treaty until today, and the coming into force of the Lisbon Treaty, the OLP has steadily increased in significance and has today taken centre stage by not being 'merely' known as co-decision, but also as the Ordinary Legislative Procedure. The OLP is, in other words, interesting for a variety of reasons and has been approached from different perspectives. What has been missing, however, is a more thorough analysis of the democratic qualities and implications of both the formal set-up and the informal practice of decision-making under OLP. This book has sought to contribute towards filling this gap by specifying more thoroughly what the democratic standards should entail, when they are met and how and why they are also relevant at a supranational European level. With the coming into force of the Lisbon Treaty, this issue has become even more pressing as the OLP now covers the majority of legal bases resulting in legislative decisions in the Union.

More specifically, the objective of this book has been to evaluate whether the procedural qualities of the OLP warrant the presumption of democratically legitimate outcomes. To this end, I have approximated deliberative democracy to empirical conditions through the development of an evaluative scheme composed of five criteria with accompanying indicators. The findings reveal (as could be expected) that there is a rather large gap between the EP and the Council with regard to how they (as individual institutional actors) comply with the criteria, i.e. that organisation and practice in the EP is more conducive to democracy than the Council's way of organising and doing things. On the basis of the findings under the various criteria – how does this affect the democratic qualities of the OLP as such?

In this last chapter, I seek to take stock of the findings from a more general perspective. I start by looking at the democratic qualities of the OLP and argue that the OLP does indeed have democratic qualities. It provides a democratic forum in the EP, openness, time allocated for debate prior to decision-making and ex post accountability mechanisms such as judicial review and Ombudsman etc. However, for the reasons outlined in the second part of this chapter, these democratic qualities are weakened due to severe democratic deficits and must therefore be conceived as potentials rather than fully realised practices of how the OLP is actually conducted. The democratic deficits discussed cover the Council's internal organisation and lead to the conclusion that, due to the heavy involvement of the working groups, the Secretariat and Coreper hinder the occurrence of a democratic opinion- and will-formation process. The second main democratic problem is the trialogues, whose scope and magnitude has seemingly become one of the procedure's most dominating features – to the extent that it is possible to speak of them as a competing decision-making structure, as argued in Chapter 9.

In the third part of this chapter, the democratic qualities and deficits discussed earlier, are summed up in an analysis of the likelihood that OLP acts – in general – meet the conditions for working agreements. Finally, the book reflects upon the evaluative framework and focuses on how the findings are related to the quest for democracy beyond the nation-state.

Democratic qualities – but not fully materialised...

As we can see from the evaluation in the previous chapters, the OLP does not fully approximate all the criteria necessary for meeting the demands of a democratically legitimate decision-making structure. This does not, however, mean that the OLP is without democratic qualities.

Inclusion of affected parties and a democratic forum

Both the EP's committee and plenary settings are organised in a way that make them conducive to democratic decision-making and hence qualify as democratic deliberative meeting places. This is important as modern democracies rely on representation rather than direct participation and the way decision-making processes are organised is crucial in order to ensure that there is a link between decision-makers and those affected. An important factor to achieve this end is the existence of a publicly open and accessible forum where elected decision-makers (or a representative selection among them) exchange, test and scrutinise arguments and positions. The evaluation revealed that the EP committees and plenary sessions are organised in such a way that it is reasonable to expect that positions and counter-positions can be tested and scrutinised in an open and accessible manner (even if there should be roll-call voting at both committee and plenary levels, as well as verbatim records of committee meetings).

Moreover, it is particularly the interplay and the division of labour between committees and plenary sessions that together meet the requirements of a democratic deliberative meeting place. On the one hand, the committees function as settings where dossiers can be discussed more thoroughly. Here the number of MEPs is reduced, but committees nevertheless ensure a representative selection of MEPs. The interaction is more informal and seemingly conducive to the exchange of (and response to) arguments. The plenary, on the other hand, is a setting with a significantly higher number of participants, which from the outset makes a thorough discussion in the deliberative sense more difficult. Hence plenary sittings are more formal and less conducive to spontaneous discussion. However, even if these institutional constraints make it possible that the MEPs themselves have already made up their mind and are, therefore, presumably not very inclined to engage in thorough deliberation with each other, the plenary is a forum – to a greater extent than committees – where the MEPs present and justify their positions towards the outside world. What is more, the EP plenary sessions are easily accessible both with regard to documents and debates (even if the default voting procedure should be roll call) and it is thus possible for interested citizens to monitor their MEPs by scrutinising the suggested amendments and speeches they make. In addition, the plenary is a setting where it may be possible for the MEPs to hold each other to account for their conduct in other (more secretive) settings such as the trialogues and the Conciliation Committee. In this sense the plenary, in particular, thus becomes a setting compatible with public justification and account-giving. In sum, the division of labour between committees and plenary sessions can be conceived as an institutional set-up ensuring a sequencing of deliberative tasks where the former performs the more detailed and explorative work and the latter is a setting where the MEPs defend their positions.

The division of labour between committee and plenary also ensures inclusion of the views of affected parties at both levels, as the MEPs are directly and substantially involved at all stages of the policy process. Unlike the Council, the process of moulding a collective will is not 'out-sourced' to unelected officials. Hence the normative hierarchy between affected and competent parties is respected in the Parliament. In this sense, the formal set-up in the EP does not only allow for deliberation prior to decision-taking, it also ensures a *democratic* opinion- and will-formation process. Even if not a democratic feature in itself, the importance of a democratic deliberative meeting place is crucial for representative democracies as it helps to build trust among those affected and their representatives.

We may add that, albeit a to varying degree, the open EP settings also help the public to get better acquainted with some of the reasons behind the Commission's and the Council's policy choices and amendments. As we have seen, whereas the Commission always participates in both the committee and plenary sessions – it explains and justifies its position in front of the MEPs – the Council is more hesitant to respond to the MEPs' questions and usually prefers to use these meetings as listening posts.

Openness

In comparison to other EU decision-making procedures, the transparency rules governing the OLP ensure a greater degree of openness. The reason for this is, first and foremost, the full participation of the EP under the OLP which, as we have seen, conducts its legislative processes in greater transparency than the other two institutions. On its part, the Council's own rules of procedure now commit the institution to open discussions at the ministerial level at all stages of the OLP (cf. Article 7). In contrast, open discussions under other procedures, for instance when dealing with non-legislative proposals, are limited to the first debates on new and important dossiers or, otherwise, if the Presidency and Coreper decide to open the ministerial meetings (cf. Article 8). Hence, discussions on non-legislative items are not automatically conducted in openness. For the public discussions at the ministerial meetings, the Council provides internet streaming, which makes it possible for everyone with a computer and internet access to follow the sessions. In addition, the Council's practice of publishing voting charts, which are attached to the legislative dossiers, ensures intelligibility of ministerial voting.

Altogether, when the current treatment of OLP dossiers is compared to the Council's openness record in the past (prior to the Brussels Presidency Conclusions and the subsequent revision and improvement of the rules of procedure in 2006 and 2009) as well as to the present rules concerning other decision-making procedures, openness appears to be considerably better under the OLP than under other EU decision-making procedures. Given that the number of legal bases falling under the OLP has been doubled with the coming into force of the Lisbon Treaty, we may also argue that this has helped to make the EU more transparent overall – at least in theory. However, and while these improvements are acknowledged and underlined as important steps in the right direction towards facilitating the formation of an informed public opinion, the sections below will further substantiate these findings in accordance with the other conclusions that can be drawn from the evaluation. More precisely, the democratic qualities of the OLP must be balanced against the democratic deficits of its actual practice.

The formal set-up of co-decision – sequencing of decision-making stages

From a deliberative democratic perspective, decisions should not be taken before a proposal has been properly discussed and examined from all relevant perspectives. Political choices that will be binding on citizens require argumentative testing. However, when positions have been presented and considered, an element of institutionalised bargaining is legitimate and acceptable in order to ensure that decisions are also taken efficiently. Bargaining and subsequently voting, are mechanisms to reject or adopt the relevant legislative proposal when time is up and the parties have not reached a consensus. Moreover, voting can also represent a formal confirmation of a process wherebye a consensus has been reached. Generally, the point is that the *sequencing* of policy-making stages is central.

When we look at the formal set-up of the OLP in Article 294, the lack of strict time limits gives the first reading an explorative character, providing both the EP and the Council with the opportunity to become familiar with the content of the proposal. It offers the possibility to discuss and reach an agreement on what is normatively and technically at stake and to collectively form their positions before a final decision is taken in each legislative house. In this sense the formal rules allow for deliberation prior to decision-making. The second reading is more structured with regard to time limits, but debate is still possible. What is more, if proceedings under the second reading are difficult but agreement is nevertheless perceived to be within reach, recourse to the Conciliation Committee opens up for a *Treaty-regulated* bargaining setting between the EP and the Council with the Commission as facilitator and broker. The reason why the Conciliation Committee can be seen as a legitimate bargaining setting is because it appears *towards the end* of the procedure when the positions and options are well known to the participants and to the public, but final agreement has not yet been achieved. Moreover, even if its members are not composed of elected representatives only, the Conciliation Committee can still be conceived as a legitimate bargaining arrangement. In this way, it provides a legitimate mechanism to induce both efficiency and effectiveness in co-decision-making. The third reading does not open for debate, but a presentation of the agreement reached in the Conciliation Committee. Hence the third reading procedure in both the EP and the Council respectively amounts to mere voting on a (reversible) piece of legislation. Again, the formal set-up of the OLP, with an explorative potential in the first reading, more structured proceedings in the second reading with time for both debate and possibly regulated inter-institutional bargaining towards the end of the process, makes a third reading solely devoted to voting acceptable.

In sum, the Treaty-regulated set-up of the OLP ensures that deliberation can precede bargaining and decision-taking. Moreover, the sequencing of the procedure allows for intelligible stages of decision-making. However, the established practices of co-decision-making, as manifested by the trialogues, obscure the democratic qualities of the formal Treaty set-up and thus, as we shall see below, render them merely potentials.

Intelligibility of procedure and separation of powers

A similar pattern can be found with regard to the ex ante indicators, described under the fourth criterion regarding mechanisms to neutralise asymmetrical power relations.

With regard to the first neutralisation indicator – intelligibility of procedure – the description of the OLP in Article 294 is, I would argue, fairly comprehensible for citizens and other interested parties. Moreover, co-decision-making is now also recognised as the ordinary legislative procedure (OLP) in the Treaties after the coming into force of the Lisbon Treaty. Earlier, this was not the case as the name 'co-decision procedure' was not written and recognised constitutionally in the Treaties (see ex Article 251 TEC; cf. Stie, 2010a). This intelligibility of the OLP is also supported by the fact that the Commission, the EP and the Council all provide easily accessible and comprehensible guides which explain the basic functioning of the procedure. Article 294 delineates a predictable and consistent procedure in the sense that it describes the three readings, its various time limits, voting rules and how it is possible to conclude a dossier at every reading. Moreover, Article 294 describes the actors involved and how the Conciliation Committee shall be organised and composed. In this sense, we may argue that the OLP is an intelligible legislative procedure.

This is also the case when we consider the principle of the separation of powers. When seen in isolation from the larger EU system, the legal demarcation between the Commission as the policy-initiator and proposer, the EP and the Council as the co-legislative houses and the ECJ as the judicial branch, is respected. However, when the larger EU inter-institutional relations and powers are under consideration, the EU system allows the Council another opportunity to look at the OLP dossiers since the same national experts often sit in both the Commission expert groups as well as the Council working groups (and later in the Comitology committees).

Hence, it can be argued again that the formal arrangement as described in the Treaties is conducive to a democratic organisation of co-decision-making, but that the established processes and practices weaken the democratic quality intended by the formal provisions.

Judicial review and Ombudsman as additional accountability and justificatory settings

With regard to the two ex post neutralisation mechanisms, we saw that both the ECJ and the Ombudsman offer settings where decision-makers can potentially be held to account.

OLP dossiers can be tried before the ECJ if (natural and/or legal) persons consider their rights to be violated. Here, the Court has jurisdiction and is authorised to review not only compliance with procedural rules, but also whether the contents of OLP dossiers are in line with Community law. As a neutralisation mechanism, the Court functions as a setting where decision-makers have to provide reasons and justify their conduct/decisions in a publicly accessible manner. Hence, knowing that they may be tried before the Court *after* an act is adopted *can* have a disciplining effect on the actions of decision-makers *during* the legislative procedure.

Even if only a soft law mechanism, the Ombudsman institution has had a significant impact on opening up decision-making processes in the EU. The EO has steadily carved out a place as an important complaint body, especially with regard to procedural aspects such as lack of openness. In such cases, plaintiffs can get their case reviewed in an easily accessible and inexpensive manner. In this sense, the EO offers a setting where individuals with limited means can have their complaints reviewed (given that the matter falls into the category of maladministration).

Both the Court and the Ombudsman can thus provide third-party scrutiny that may intervene between decision-makers and those affected when the latter consider their rights to be violated by OLP dossiers. Both are procedural mechanisms that *can* function as limitations and incentives on decision-makers' actions, so that they are more likely to behave in accordance with Treaty provisions. In this sense, the Court and the EO can have pre-emptive effect. However, even if the potential is there and the Court has shown that it can function as a power neutralising mechanism in this regard, it should be added that the actual number of individual plaintiffs before the Court is rather small.

In addition, and equally important from a deliberative democratic perspective, both the Court and the Ombudsman represent publicly accessible settings where decision-makers must explain and justify their positions. We could also stretch this point further and argue that, even in cases where the plaintiffs do not succeed, the proceedings of the Court as well as the Ombudsman put the burden of argument and justification on the decision-makers rather than on the plaintiffs. They ensure that single individuals and/or other legal persons do not have to stand up against the EU institutions directly, but are rather helped by an adjudicative third-party. Such mechanisms are important, as a direct address by individuals and others can be very easy to simply ignore at the EU level due to the very asymmetrical relationship between the OLP institutions and single individuals and/or other legal persons. This is especially so in polities without a well-functioning media structure and public sphere, such as the EU.

Democratic deficits

These democratic qualities notwithstanding, the OLP's overall problem is that the above-mentioned qualities are seriously weakened due to the character and scope of informally established rules and practices. This pertains first and foremost to the extensive usage of the trialogues, but also relates to how the Council is organised internally. More precisely, whereas the above shows that the formal set-up of the OLP clearly has democratic qualities, the way decision-making under the OLP is actually conducted and practiced largely renders these qualities latent rather than materialised.

The Council's internal organisation – an obstacle to a democratic opinion- and will-formation process

It was argued above that, in order for the OLP to fulfil the first criterion, both legislative houses must ensure an open setting conducive to the requirements of a democratic deliberative meeting place. As we have seen, the Council does not live up to its role as a democratic legislator according to this understanding. The problem is that it is primarily the lower echelons – the working groups, the Council Secretariat and Coreper – that take care of the daily work on OLP dossiers. Hence the important collective opinion- and will-formation process prior to actual decision-taking is not in the hands of the people and their (elected) ministers, but rather those of the unelected Council officials. It may be that the agreements accomplished in these settings are reached in a deliberative manner, but they do not have democratic qualities and can only qualify for the category of deliberative legitimacy as introduced in Chapter 2. At the ministerial level, on the other hand, the process is better described as an aggregation of preferences rather than a collective forming of a Common Position. In this way, Union citizens are deprived of a setting where arguments are presented, tested and justified in a publicly accessible manner. The meetings below the ministerial level are all closed and conducted in camera. Apart from provisional agendas, there are no working documents available from these meetings and it is only the Commission that has access to Council meetings. The EP and the public, on the other hand, are generally excluded from knowing what (and how) dossiers are discussed in the working groups and in Coreper.

Consequently, the heavy influence of unelected officials renders the openness of ministerial meetings de facto less valuable from a democratic point of view. Open ministerial meetings function at best as settings where citizens can view the adoption of already reached agreements. In this way, the institutional interplay and division of labour between working groups/Secretariat/Coreper, on the one hand, and the ministerial setting, on the other, does not comply with the conditions for a *democratic* sequencing of deliberative tasks/moments as is the case in the EP, since the Council officials have no authorisation to speak on behalf of affected parties. It is only the ministers who have the electoral foundation to be the citizens' spokespersons, but in leaving almost the entire legislative process in the hands of the unelected officials, they fail to live up to the requirement of public justification. As argued in this book, it is not enough merely to be elected as representatives, they must also earn their right to speak on behalf of citizens. This is only achieved to the extent that they are willing to explain and justify their decisions in a setting where the public can actually watch them 'in action'. In other words, Council ministers are not in line with the requirement in the Dag Hammarskjöld slogan 'Only he deserves power who every day justifies it', as the exercise of power is not satisfactorily fused with public justification either at the EU level or at the national level. The Council's main challenge is, therefore, that there is no *democratic* process of collective will-formation, only at best a democratic voting or decision-taking moment:

> Unlike most other legislatures, the Council members do not pass policies according to a set of policy preferences publicly announced to the full population affected by these policies. It is thus often difficult for the constituencies to have a clear overview of all governments' positions on a specific policy issue.
>
> (Hagemann and De Clerck-Sachsse, 2007: 40)

There is much talk about the fragile status of the EP in the hearts and minds of European citizens, but they should perhaps have the same worries about the Council. It is difficult to ignore the fact that the internal organisation of the Council suffers from serious democratic deficits, which consequently also affect the democratic quality of the OLP. The indirect chain of representation is from the outset a weak democratic mandate. To function as a democratic legislator, the Council should consequently be organised in a way that aims at compensating for this shortcoming, not the other way around where it shuns publicity and its obligation to justify decisions. The ministers do not even meet the legitimacy requirement on which the intergovernmental logic rests. Here, the ministers should at least comply with what the principle of indirect representation demands and have established a practice of reason- and account-giving in their respective constituencies at the national level. Instead, citizens are merely cut off with a press conference or statement towards the end of the process and this does not comply with even the minimum democratic requirements. The implication of the intergovernmental logic underpinning the Council's organisational set-up effectively distorts the legitimacy of the ministers' indirect chain of representation:

> Secrecy creates and reinforces the informational and capacity asymmetries which plague the relationship between national executives and national parliaments on European issues. It precludes national parliaments from having sources of information alternative to the accounts offered by national ministers ... This renders it difficult not only to know what has actually been said in such meetings, but especially *why* decisions were adopted.
> (Menéndez, 2009: 303; see also Schneider and Baltz, 2005: 23)

In fact, the very basis on which the intergovernmental principle of indirect representation rests can itself be questioned. Many studies have shown that the Council is not so intergovernmental after all, but rather a highly integrated and supranational institution.[1] Hence the principle of indirect representation is not enough to legitimise the ministers' delegated powers and this, subsequently, also renders the intergovernmental evaluation standard insufficient. This is not only detrimental for the supranational legislative process (which we have been dealing with in this book). As mentioned above, it also seriously affects the national level and the national parliaments' ability to conduct an informed and enlightened scrutiny of their government.

This argument still holds, I would argue, even though the national parliaments now have the right to intervene in the legislative process through the aforementioned early warning system if they find a proposal in violation of the principle of subsidiarity. It is doubtful that the yellow or orange card procedures possess enough clout to function as scrutinising mechanisms of the national governments when they act in the Council. Several reasons could be mentioned. First, is the blunt fact that the early warning mechanism is narrow in scope as it can only be invoked in subsidiarity cases. The early warning mechanism could perhaps be seen as a supplement to the monitoring of the national parliaments' conduct through their respective European affairs committees. The quality and influence these committees exercise are, however, variable, ranging from the quite active Danish European affairs committee to some of the Central and Eastern European member states' committees which are hardly active at all (cf. Buzogány and Stuchlík, 2011). Second, it requires both the willingness and ability of national parliaments to cooperate and act collectively in order to drum up the necessary majorities to even start a yellow or orange card procedure. Whereas collective action by the national parliaments should not be ruled out as such, the prospect of coordinated action is demanding especially given the fact that EU legislation is not the main concern of national parliaments.[2] Third, and seen from a more supranational perspective, given that the EP and the Council both have an electoral anchoring (even if only indirectly with regard to the Council), it could also be argued that it does not seem necessary, from a democratic point of view, to have more parliamentary bodies involved during the legislative process. More actors could potentially obstruct the process and lead to inefficiency. It is well known that the time dimension is critical and that the national parliaments are already pressed to find time to perform a meaningful control function of EU policy-making (see Lord, 2004: 178–82). In addition, and as argued above, it is an important democratic virtue to ensure intelligibility and transparency of the decision-making procedure and too many actors can violate this principle. Fourth, it is argued that the early warning mechanism is not meant to be triggered on a regular basis, but should be seen as a right to be invoked in those (hopefully seldom) instances where the national parliaments find that an OLP proposal is violating the subsidiarity principle (cf. Kiiver, 2011). Last, but not least, the early warning mechanism has, so far, been filed only once and this raises the question of the importance that national parliaments actually attach to this right and thus how effective it can be as a means to compensate for the Council's unwillingness to act as a true democratic legislator, honouring its duty to justify policy-making publicly towards Union citizens.

Politicians and experts

Some may find the conclusion on the internal organisation of the Council too harsh. They may side with intergovernmental logic, where ministers work on delegated powers and where they are only accountable in their respective national constituencies. In this regard, it should be added that the second criterion of inclusion of affected and competent parties reflects an attempt to capture empirically the difference between the democratic and the epistemic dimensions of decision-making. In arguing that affected parties' representatives should have the upper hand, critics may argue that the claim I make here is too strong. Modern decision-making is complex, technical and knowledge-demanding. Elected representatives are not experts in all fields (even if the sectoral division of EP committees has fostered many quite specialised MEPs), but are politicians with general (rather than specialist) knowledge.

In expert terms, then, democratically elected legislators are considered:

> to be seriously deficient as lawmaking bodies. They often lack, among other things, the expertise, staffing resources, organisation, cohesion and flexibility needed for the making of quality legislation. Thus, in no European country does the elected body act as the exclusive legislator.
>
> (Van Schendelen and Scully, 2006: 2)

In other words, experts in a variety of fields are needed to ensure that decisions meet certain scientific and cognitive demands. Numerous experts are included and consulted in many ways, ranging from single events with outside lobbyists to permanent sector bureaucrats in the Commission or the national experts in the Council working groups. With regard to the internal organisation of the Council, the working groups, the Secretariat and Coreper indeed contribute to much needed (efficient) problem-solving, technical expertise and the ability to reach agreement among a heterogeneous set of member state positions. Hence they also play an important role in co-ordinating collective action and in enhancing the epistemic dimension of OLP dossiers, which are also valuable to democratic decision-making.

However, from a democratic perspective the fear is that they become too powerful thus out-manoeuvring the elected generalists. As discussed above, this is a challenge in the Council. Here, the ministers play second violin to working group, Secretariat and Coreper officials in the important opinion- and will-formation process prior to final decision-taking. The latter run the 'daily show' (largely due to the fact that sitting in the Council is only a part-time job for the national ministers) and they also influence the way issues are formulated and discussed. As we have seen, issues are framed in technical terms which are subsequently adopted at the political level where ethical and moral dilemmas, political choices and conflicting positions should be clearly articulated and communicated to the general public in order to bring out the normative implications of policy proposals. This type of technical language sometimes also influences debates in the EP. However, OLP dossiers are not merely apolitical and uncontroversial, and the application of a technical language helps to 'neutralise' and 'disguise' both the political salience and the stakes from citizens and other affected parties.[3] The plea for more early agreements has partly been justified on the basis that there are many dossiers that are merely technical and can thus be more easily and efficiently dealt with if they are fast-tracked through in a first reading where they – from the Council side – are unlikely to be managed by an elected representative.

The fear of expert domination is not unique to the EU alone even if it is more critical at the European level, due to weaker representative institutions, to ensure democratic control. However, the Council's problem is a general one, namely that:

> officials act on the basis of normative knowledge that is not regulated by democratic procedures. They use their 'own' perceptions and values or 'internal' professional standard when they evaluate societal problems. The question is from where the experts yield authorisation to deploy particular values when doing their job. Who has given them the right to act in this way and how can we know that they have applied correct values and valid knowledge?
>
> (Eriksen, 2001: 17–18, my translation)

In other words, the chain of democratically legitimate decision-making has been turned upside down here, as administrative power has not been democratically assessed and transformed into communicative power before the Council decides either to adopt or to reject an OLP dossier (cf. Chapter 2). The internal organisation of the Council hinders OLP dossiers from being scrutinised by public reason because the balance between the politicians and experts tilts towards the latter to the extent that the fear of a self-programming 'epistocracy' or 'governance without democracy' is imminent (Eriksen, 2009: 173–4). This is also underlined by the fact that there are no procedures to hold these officials accountable either through elections or in public debate. Hence the scope and extension of the Council officials' involvement should not be mistaken as a democratically legitimate way of relieving the ministers of their heavy workload, as the arrangement seriously violates the normative hierarchy between elected and non-elected actors.

From the outset, the strong emphasis on respecting the normative hierarchy between representatives of affected and competent parties may seem a bit rigid. True, the consequence of upholding this emphasis is that in cases of conflict the views of elected politicians will always have to trump those of the experts, thus also running the risk that the representatives of affected parties' views may not offer the qualitatively best choices and therefore the most optimal outcomes for all those affected. This is of course *not* the desired outcome, but it is arguably the price that must be paid if we want to retain a certain level of democratic control of decision-making. Since we cannot anticipate all the consequences of a decision, we can never really be sure about what the best solution is. No one possesses complete information and knowledge – not even experts. Unintended consequences must always be factored in – sometimes they go in a positive direction, but very often also in a negative direction. Unelected and unaccountable experts cannot, and should not be, allowed to dominate the majority of the opinion- and will-formation process or have the last decision-taking word.

It is, however, impossible to fully avoid a situation where experts deal with issues that raise ethical and moral questions, but it is the way and the extent to which they are involved in the Council decision-making process that is worrying from a democratic perspective. However, to decide exactly how the balance between politicians and bureaucrats/experts should be drawn is difficult and is ultimately, a task to be achieved through public debate between Union citizens and politicians. In other words, if the Council insists on keeping its secretive working methods and privileges, then ministers should be prepared to defend and sustain them in a public debate.

Moreover, the democratic deficits of the internal organisation of the Council do not prevent dossiers from being tested argumentatively. Many studies show that the working groups and Coreper are settings where decisions are reached in a more deliberative way (Lewis, 1998, 2000; Heisenberg, 2005; Hayes-Renshaw and Wallace, 2006). The point is that this type of deliberation does not amount to a democratic test of arguments as the ministers are not involved and the meetings are not publicly open. Whereas the more deliberative and consensus-building method characterising the lower level working groups in the Council seemingly represents a strong candidate for what I labelled *deliberative legitimacy* in the introductory chapters, the evaluative scheme and the subsequent assessment above clearly illustrate that this working method does not meet the requirement of democratic deliberation and thus should not be confused with *democratic legitimacy*.

Having said this, the normative aspects and implications of experts' and bureaucrats' involvement in EU decision-making is, on the whole, still an underdeveloped research area that should be investigated and assessed more thoroughly. For instance, they do not, as such, have a 'constituency' to which their knowledge is attached, that is, they advocate and exercise theoretical and skill-based knowledge, which is acquired through several years of education and training. On this basis, they could argue that their deliberations are mainly scientific and not political; that they decide within given parameters; that they are constrained by mandate, procedural rules and organisational setting; and that they are controlled by other colleagues. Hence, even if the participation of experts and bureaucrats cannot, from a democratic perspective, dominate the decision-making process, the professional knowledge with which they contribute represents – in principle – a more general or universally valid kind that does not necessarily reflect and represent parochial or special interests situated in civil society, the market or the political system as such. It could (perhaps) rather be argued that their participation has been an important component for the development and stability of the constitutional state and that, in this sense, they represent important models of democratic values and principles. The more precise normative value of how experts and bureaucrats can (and are) contributing democratically in general, and in the EU in particular, cannot however, be decided in advance.

The democratic challenge of trialogues

It is not only the internal organisation of the Council that affects the OLP's democratic credentials. Throughout the book, the impact of trialogues has been a topic in almost every chapter and it is now time to address them from a more general perspective. In short, the problem with how the trialogue system functions today, is related to its scope and the extent to which it now influences the OLP processes. This results in the following democratic problems:

1. Trialogues interrupt the collective opinion- and will-formation processes in the EP.
2. They keep crucial parts of the decision-making process away from the public eye and thus reduce the level of openness.
3. They reduce the number of decision-makers.
4. They are not composed of a representative selection of elected participants, hence citizens cannot be sure that all positions have been voiced and taken into consideration.
5. They make the OLP less intelligible for citizens as the emphasis is on informal and secret meetings.

6 They make it harder for citizens to know what decision-makers are deciding and hence also more difficult to hold them accountable.
7 They (may) weaken the quality of dossiers.

One of the democratic qualities of the OLP reported above was that the division of labour inside the EP does not violate the normative hierarchy between elected and unelected actors since, first, the MEPs are involved in all stages of the decision-making process and second, the EP committees are representative of the EP as a whole. Moreover, the division of labour could be perceived as a sequencing of deliberative moments/tasks ensuring a democratic opinion- and will-formation process in the EP. However, the evaluation also revealed that this internal process between committee and plenary is to a significant extent *disconnected* due to the constant pace of trialogue meetings. When committees and plenary sessions become less important this also deprives Union citizens and other affected parties of their right to justification. Thus, trialogues represent a serious obstacle to the *democratic* process of opinion- and will-formation process taking place in Parliament and hence, to a large extent, disqualify the democratic features of the formal OLP process discussed earlier. What is more, the impact of trialogues is exacerbated when agreement is reached in the first reading. The process is, as we have seen, even more informal at this stage and is impermeable to citizens and other affected parties because at this point, the institutions have not reached an official position. In this way, the trialogues circumvent the formal set-up of the EP, which (in principle at least) ensures that there is one democratic deliberative meeting place where decision-makers defend their positions on a regular basis. However, the established practice of co-decision-making is, in reality, that the scope and scale of trialogues keeps the majority of important and decisive legislative discussions more or less constantly outside the public radar.

Trialogues also harm the criterion of inclusion, as there are only a handful of people present and neither the Council nor the EP participants constitute a representative selection of their respective houses. Rather, in the EP, we have seen that the larger political groups have become more dominant as they often occupy the key positions (rapporteurs, committee chair, group coordinators) and thus become decisive actors. In the Council, delegates from the country holding the Presidency dominate (i.e. officials from the working groups, Coreper and, potentially, the minister). From a deliberative perspective, the problem with closed settings under such circumstances is that the deliberation that may take place might be too narrow and fail to 'reproduce the pluralism of the public in the private' (Chambers, 2004a: 390). Trialogues risk being dominated by 'private reasons' instead of 'public reasons', which means that the arguments presented are not arguments that all could generally accept if they were presented in a publicly open debate (i.e. egoistic and self-interested reasons).

The problem is not necessarily that trialogues contain a limited number of participants. This could be legitimate if they were representative of the interests and positions in the EP and the Council respectively and/or the results of trialogue deals were properly tested in a publicly accessible manner after the meeting had taken place and before a final decision was made. Instead, the composition of trialogue members challenges the principle of representation as we cannot be sure that all positions have been taken into consideration when, first, there is not a representative selection of MEPs and, second, when the Council is not represented by elected representatives. And with regard to the democratic testing of the trialogue outcomes, the assessment showed that these outcomes are often presented to the EP committees and plenary sessions as 'take-it-or-leave-it' deals rather than as starting points or bases for a democratic legislative process. As for the Council, the trialogue participants (that is, the permanent representative from the member state holding the Presidency) usually orient and consult the permanent representatives and not the ministers concerning the acceptability of a trialogue outcome. How can Union citizens trust that their positions have been properly voiced and considered under such circumstances?

Whereas the interaction between the EP and the Council has increased significantly (at all levels) with the introduction of the OLP (Corbett *et al.*, 2007: 242) and has also improved inter-institutional relations, this has seemingly happened at the expense of the intelligibility of the procedure. As we have seen, the OLP has been accused of being a complicated procedure, where the protracted process of informal inter-institutional interaction through trialogues makes it very hard for the public to grasp what is going on during the process (see Bostock, 2002; Andeweg, 2007; Bunyan, 2007b, 2007a: 2; Fouilleux *et al.*, 2005: 617). The inclusion of elected representatives cannot compensate for the insulated and non-transparent procedural features because the latter deprives the citizens of the chance to hold co-decision-makers to account. In this sense, the trialogue system is not a procedural feature that induces, but rather deters public reason-giving towards affected parties.

What is more, the trialogues have seemingly also made the OLP more bureaucratised, encumbering political input as the focus is on reaching deals as quickly as possible before presenting them to the EP and the Council for approval. Whereas the institutions have done a lot to improve the relations between them, such as coordinating calendars and working methods and streamlining the functioning of the trialogues in the Joint Declarations, their relationship to Union citizens has been neglected. It could thus be argued that the trialogues make the EP and the Council more inward-looking, minding their own deals and discussions and 'forgetting' about the public's right to know what they are up to. The result is that the procedure could be perceived as more of an inter-institutional battle between the EP, the Council and the Commission, rather than a political battle over the best course of action because the citizens may find it hard to detect what the political stakes really are. In other words, the institutions' duty to identify and articulate politically salient issues and dilemmas in an open and easily understandable manner suffers. This is especially so concerning the commitment to close dossiers as quickly as possible in early agreements. The Joint Declarations underline efficiency and transparency, but whereas there are many paragraphs on how to ensure efficiency, the transparency and accountability aspects are merely mentioned, not developed. The unilateral introduction of the Code in the EP where trialogue-MEPs, among other things, are encouraged to seek mandate from the committee as a whole and to avoid closing deals before they have been discussed in committee is an effort to make the process more accountable. The democratic impact of the Code remains to be seen, however, even if Toshkov and Rasmussen (2012) indicate that it may have influenced positively since the duration of the first reading is longer when the institutions deal with politically salient dossiers, and these are closed as early agreements rather than after the second or third readings (I come back to this point below).

Even if the Joint Declarations semi-formalise the trialogues to some extent, the Council and also some MEPs (mainly from the large political groups who benefit from current arrangements) seem to be unwilling to overhaul and formalise trialogue rules and practices through an Intergovernmental Conference. They want to keep them as informal and flexible as possible because this enhances the opportunities for the key actors to influence the dossiers. What is more, there seems to be a general tendency within the EU at large to by-pass the formal procedure for Treaty reform and use inter-institutional agreements (such as the Joint Declarations) to alter the formal, institutional set-up in a favoured direction even if this is not in line with Treaty provisions (cf. Curtin, 2007a). Without the authorisation of an Intergovernmental Conference, the development and semi-formalisation of the trialogue system through the Joint Declarations has changed the character of the OLP from a sequential to a simultaneous process which, in effect, is violating the current Treaty provisions as set up in Article 294. In turn, this has resulted in the situation that the trialogue system has become a competing decision-making arrangement, which largely empties the formal processes in the EP and the Council of legislative importance.

There is yet another (though related) problem regarding the extensive use of trialogues, which relates to how they affect the overall quality of OLP acts. Whereas the trialogue system plays an important part in explaining the successful adoption of OLP dossiers, the efficiency gained may backfire on the whole system and challenge the democratic improvement represented by the increased involvement and competence of the EP in the Union's overall decision-making. The reason for this again, first and foremost, pertains to the negative effect that early agreements have on the democratic quality of OLP acts. Prior to Amsterdam, early agreement was not an option and most dossiers went all the way through conciliation which meant that the institutions' positions and arguments were better known, first, because the EP and the Council had by then actually formed a position and, second, because after the first reading more documents are accessible (i.e. the EP's Position, the Council's Common Position etc.). Today, very few dossiers reach the conciliation stage, and even in a situation where the 2004–2007 enlargements significantly increased the number of participants and presumably also made it more difficult to reach agreement, the trend is still towards first reading agreements.

However:

> although more readings are not required before agreements can be made, it was at many occasions stressed that the meetings last longer, and that the legislative process may have been prolonged somewhat with regard to the time spent – rather than the number of readings – on each individual proposal.
>
> (Hagemann and De Clerck-Sachsse, 2007: 25)

This tendency is also illustrated by the EP (2009b: 12), 'It is interesting to note that the average length of codecision procedures has been only very moderately shorter during this legislature [i.e. the sixth parliamentary term] than during the 5th term despite the increase in early agreements.' In other words, even if dossiers are concluded in first readings, it should not automatically be concluded that the decision-making process is always much more efficient if a dossier is closed at first rather than at second reading. The reason for this is, as we have seen, that there are no time limits in the first reading. The process can continue as long as the institutions so decide. Even if the argument is that early agreements are preferable because they contribute to efficient legislative processes, it has also been shown that the reason why the institutions prefer (and are in the majority of cases also able) to close the process after the first reading, is that at this point, the decision-making situation is more flexible (though also more undemocratic). Or as the EP (2009b: 13) states:

> This trend [that the average length before dossiers are adopted has increased] can be explained with the higher number of 1st reading agreements which include more and more controversial files which need time to be negotiated and where Parliament strives to get the best deal.

That the duration of first readings is longer when acts are adopted as early agreements rather than in second or third readings, is also corroborated in Toshkov and Rasmussen's (2012) study. This trend was found when the EP and Council were to decide on highly salient dossiers. On this basis, Toshkov and Rasmussen (2012: 17) conclude that:

> these findings do indicate that the co-legislators compensate for some of the lost time from not going to second reading by extending the length of the first reading negotiations when the deals are salient. Rather than finding evidence that deals are pushed through quickly irrespective of how controversial they are, we find that the co-legislators spend more time on the first reading negotiation period the greater the level of political disagreement between them.

However, the question is whether the extra time that is spent during the first reading, when dossiers are closed as early agreements, ensures that debates are actually subject to more openness. Even if the EP's Code may induce more openness, it is not binding on the MEPs and there is consequently no guarantee that longer first readings can be equated with more transparency. Rather, the absence of official EP and Council positions which function as a constraining factor on the key actors' scope of independent manoeuvring in the second reading is not as strong in the first reading. Hence, the flexibility of the first reading is – as we have seen – also among the reasons why the use of trialogues so early in the decision-making process is severely undermining the conditions for democratic legitimacy at this stage.

Finally, it should also be questioned whether reaching a deal early in the process among only a handful of people who meet in secrecy, is actually a productive method of ensuring qualitatively good decisions. The extensive use of trialogues may backfire as the system becomes less open to new ideas and risks the recycling of opinions and positions which (in the longer run) can affect and subsequently jeopardise the quality of (and trust in) the legitimacy of EU legislative acts. What seems to be forgotten when urging for early agreements is that public policy-making must be able to master the difficult task of simultaneously balancing efficiency and democratic legitimacy. Hence there are good reasons for arguing that the EU legislative processes can gain from striking a better balance between these two standards and taking time to go through the formal stages of decision-making in a way that can also be followed by citizens, instead of fast-tracking legislation through the EP and the Council ministerial settings as quickly as possible. In sum, a democratically legitimate procedure may turn out to be more efficient in the longer run because it produces qualitatively better and thus more stable outcomes.

OLP acts as working agreements?

The main conclusions that can be drawn from the analysis are two-fold and are closely inter-linked. On the one hand, the formal rules of co-decision-making are in many respects compatible with democratic decision-making. Had the Council's internal organisation been improved, then the procedure could have represented a promising example of democracy beyond the nation-state. On the other hand, the established rules and practices manifested in the trialogue system render passive the democratic potentials characterising the formal procedural rules. When the main, or most important, part of the decision-making process takes place in trialogues and not in the formal institutions, the OLP becomes 'committee governance' and loses its kernel of democratic government.

Based on the above evaluation, it cannot, in other words, be argued that OLP acts are likely to meet the quality of a working agreement. As shown in Chapter 2, a working agreement is an outcome that is supported by reasonable reasons, that is, reasons that are 'inter-subjectively justifiable', they are different, but 'become understandable and mutually acceptable' through deliberative processes (Eriksen, 2007a: 107, 108). Moreover, reasonable reasons are those that have been tested and have withstood critique in a publicly accessible manner, and the only way to determine whether a decision-making procedure comes close to complying with this standard is if it meets certain procedural criteria such as those developed in this book. As we have seen, the procedural conditions underpinning co-decision-making do not ensure a proper *democratic* testing of dossiers before they are adopted. Rather, the process of co-decision-making is largely held behind closed doors and/or conducted by a handful of people – some elected, but also a significant number of unelected actors. These actors cannot be assumed to represent the plurality of views that characterise the EU. This is clearly the case under early agreements, but second reading agreements also suffer from the extensive use of trialogues. Under these conditions it is consequently unlikely to be assumed that OLP acts are supported by mutually acceptable reasons.

But maybe we are judging the OLP too harshly? In other words, is the working agreement too demanding a standard for a supranational decision-making procedure such as the OLP? In this regard we could reiterate the rhetorical question posed in the first chapter: if Union citizens were asked to actually give their consent to OLP acts, what kind of conditions must the procedure ensure in order to make it likely that citizens will find such acts acceptable and thus legitimate constraints on their actions? What reasons could they have to obey OLP acts? Most likely they will not all agree on one single democratic principle, but rather invoke different, but equally valid understandings of how the OLP should be democratically organised. Some would perhaps argue that the OLP should be organised according to an understanding of democracy as a popular sovereignty based on the principle of one person, one vote. Others would defend the idea that democracy is defined according to how convincingly political organs have justified their decisions towards affected parties. Others again would defend a principle of delegated powers, where certain political organs have obtained competence to act on behalf of the collective such as the arrangement of national representation of member states in the Council. All these arguments could in themselves be perceived to be reasonable. Seen from this vantage point, the working agreement seems like a suitable standard for assessing the OLP as it can accommodate all the above interpretations of EU democracy *at the same time* –though only as long as the various conceptions of democracy have themselves been discussed and vindicated as mutually justifiable in a democratic process.

Since its inception with the Maastricht Treaty, and up to the current Lisbon provisions, the OLP has become more important both with regard to the expansion of legal bases as well as the overall increase in the number of acts taken under this legislative framework. The OLP now covers a number of policy areas, which have hitherto been perceived to be too sensitive to allow the EP considerable influence over, such as agriculture, fisheries and various areas within justice and home affairs. The growing number of policy areas has not decreased the inclinations to close dossiers after the first reading in order to increase efficiency. Under the work of the Convention on the Future of Europe, it was argued that the OLP 'was proving successful in reinforcing the legitimacy of European legislation and should become the normal way EU laws are adopted' (Corbett *et al.*, 2007: 220). Even if the OLP has an aura of democratic legitimacy due to the involvement and participation of the EP, we should urge a bit of caution against such a presentation of the procedure. The present evaluation has revealed that this picture must be balanced against its democratic deficits. To put it bluntly and provocatively, we could argue that the way the OLP functions is sometimes more precisely described as a technocratic procedure in democratic disguise. This is not a promising prospect for democracy beyond the nation-state.

The evaluative framework and the way ahead

In this study, the focus has been on discussing and assessing the democratic qualities and shortcomings of the OLP. In this sense, the present study fills a gap in the academic literature on the OLP which has first and foremost been concerned with the issues of how the OLP affects the relative inter-institutional power relations, how to explain institutional change and/or how to understand the general developments of co-decision-making. Whereas one may argue that these debates indirectly touch upon democratic issues, for instance by investigating the status of the EP, they do not provide an analytical framework that facilitates a discussion on the democratic credentials or shortcomings of the OLP. In other words, these studies do not answer the question of how we can know democratic government when we see it. The evaluative scheme developed in this book both provides a standard (i.e. a working agreement) of what should be expected of a democratic outcome and a description of how to achieve and/or 'measure' whether a decision-making procedure complies or falls short of meeting the standard (i.e. the criteria and the indicators).[4] In this sense, the book helps to make the academic debate on the democratic merits and shortcomings of co-decision-making more tangible. It could also be argued that it provides a more concrete answer to Union citizens on when, and why, there are good reasons and when, and why, there are less good reasons to obey and respect EU law. In sum, the book helps not only to make the debate on the OLP richer; it also makes it more nuanced.

We now know how the OLP 'scores' with regard to democratic procedural qualities, but how well did the evaluative scheme itself perform? As argued above, an important aim of this study is not only to assess the OLP, but also to contribute towards the 'operationalisation' of deliberative democracy. In the final paragraphs, the task is therefore to consider the merits and shortcomings of the evaluative scheme. What can be learned from looking at procedural qualities? And where do we go from here?

First, the evaluative scheme shows that deliberative democracy is neither too utopian nor too abstract. The criteria were all operationalised into indicators that are clearly demanding but, under favourable empirical conditions, are not impossible to realise given that there is political will. What is more, both in the process of developing the evaluative scheme and in the following assessment, I have sought to emphasise that deliberative democracy is not merely about deliberation, but is rather to be seen as complementing other decision-making mechanisms (such as bargaining and voting). However, even if deliberation is not a sufficient mechanism to ensure democratic decision-making, without it, bargaining and voting will not function in a democratically defensible manner (see also Warren, 2007: 287–8). In sum, 'the proper idea of feasibility ought to be a normative one: it entails neither the surrender of the original ideals of deliberative democracy nor the mere accommodation to existing institutional and social facts' (Bohman, 1998: 423). Even if readers may disagree with the way I have developed and operationalised deliberative democracy, the argument that this model of democracy is unfeasible cannot, consequently, be upheld (I come back to this below).

Second, the main assumption underpinning this study is that democratic legitimacy is procedural and determined by the way decisions are reached. This does not mean that the quality of outcomes cannot be evaluated independently of the procedure. Outcomes can be assessed for how just or unjust, effective or ineffective etc. they are (i.e. according to other standards), but their *democratic* quality can only be determined by assessing the conditions under and the processes through which they were achieved. As argued in the introductory chapters, many EU studies have focused on the deliberative aspects of EU decision-making, but this does not mean that the normative quality we attach to agreements or practices characterised by deliberation is always the same. My argument was that without distinguishing properly between the different types of conditions under which deliberation takes place, we may wrongly overestimate the normative credentials of deliberatively reached agreements and mistakenly conclude that they automatically also have democratic quality. In underlining the difference between the role and participation of actors representing the views of affected parties (elected politicians) versus actors representing competent parties (experts, bureaucrats) and accompanied by requirements of publicity, the analytical framework is capable of conceptualising more clearly the conditions under which it is possible to expect democratic deliberation, rather than merely elite/expert deliberation, to occur.

Moreover, the evaluative framework is also equipped to show how and why it is not enough to simply include elected representatives, but that it is the level and type of involvement of these representatives that matters: 'Only he deserves power who daily justifies it'. In other words, we cannot take for granted that just because the EP was included as a co-legislator on a par with the Council, this will automatically make EU decision-making more democratic. Including the EP is not a guarantee in itself that the MEPs will behave democratically. Furthermore, we cannot simply assume that the ministers meeting in the Council can be conceived as channels where the views of affected parties (i.e. their respective national constituencies) are voiced. The evaluative framework also provides the conditions for showing when OLP acts are satisfactorily subjected to democratic scrutiny and when unelected participants become too dominant and autonomous to be compatible with democratic decision-making.

Third, a central topic in this book has been the relationship between the more open, formal set-up of the procedure versus the closed, informal (or semi-formal) trialogue structure. The evaluative framework brings to the forefront the normative dilemma inherent in this dualism. It takes us some way towards specifying the conditions under which closed settings can be legitimate and when they are violating democratic principles. We have seen that even though the Conciliation Committee did not fulfil the conditions for a democratic deliberative forum, it could be conceived as legitimate despite the fact that it is closed to the public during session. The reason for this is because it provides openness ex post, because it appears towards the end of the procedure where positions are already known and because the EP's delegation is (more or less) composed of a representative selection of Parliament as a whole. The weakness of the Conciliation Committee is consequently the Council delegation, as only the Presidency is represented at the minister level. The point to be made here is that the evaluative framework provides us with some tools to determine when a closed setting, such as the Conciliation Committee, is legitimate. The same tools also helped us to evaluate the trialogues, where it was shown that the procedural qualities are largely incompatible with democratic decision-making.

However, even if the trialogues fail in democratic terms, their existence and salience in co-decision-making indirectly raises an important issue concerning the apparent need for more sheltered settings where decision-makers can speak more freely. There is a general perception by many of those involved in (as well as observers of) co-decision-making, that trialogues are vital in order to reach agreement. More precisely, it is pointed out that the advantage of trialogues lies in the institutional format of the meetings: attendance is limited and the participants often know each other quite well, meetings are held in secret, no minutes or verbatim records are taken and hardly any information leaks out to the public prior to or during the process. Trialogues function as flexible settings where actors can speak freely and honestly, where they can sound out differences and test alternatives without being held to account immediately after the meeting is over. In this sense, the trialogues induce trust among the participants, which helps the process of reaching agreement. These qualities, it is argued, contribute towards facilitating agreement and thus towards the successful adoption of OLP dossiers.

The issue of how to balance the relationship between open and closed settings (and the alleged need for the latter), links into the broader debate on the place of publicity in democratic decision-making. This debate is also important for deliberative democracy (Elster, 1998: b and c; Chambers 2004: a and b, 2005). Moreover, and as mentioned in Chapter 2, it is often pointed out that deliberation is fragile and difficult to achieve. One important reason for this is exactly because it is dependent upon a certain level of trust and confidence. In order for deliberation to function, the actors must dare to believe that their counterparts conduct themselves truthfully and are genuinely interested in reaching agreements where the best argument is the decisive factor. In short, 'Trust is both the pre-requisite for deliberation and the result of deliberation' (Eriksen, 2003: 208). From a deliberative democratic perspective, deliberation is also assumed to ensure rationality and quality of decisions and many argue that publicity is not necessarily or always compatible with this aim (Elster, 1998c; Chambers, 2004a, 2005).

Some empirical studies have shown that open settings often have a negative influence on the quality of deliberation (Naurin, 2007; Checkel, 2001).[5] The argument is that publicity seems to trigger rhetoric, demagoguery and manipulation and the rationality of arguments thus suffers. Chambers (2004a, 2005) labels this 'plebiscitory reason'. In contrast, closed arenas often invoke the opposite, as it is easier for participants to respond to the content and quality of each other's arguments when they are not constantly confronted with media pressure and public opinion polls. It thus seems pertinent to be attentive to the conditions under which trust and confidence can be nourished and protected.[6] In a complex and heterogeneous polity such as the EU, where a common political identity (akin to those characterising national settings) is (still) lacking, trust, especially, is found wanting. On this basis, the trialogue system could be conceived as a valuable means to ensure trust-building among the decision-makers.[7] Hence, perhaps this creates a legitimate need for closed settings as they can provide the conditions for trust-building and thus more efficient (speed), effective (ability to reach agreements) and rational (ensuring quality of) problem-solving which are, after all, also important for a functioning democracy. It is often argued that closed settings are so successful (and flexible) because they, according to involved actors, are informal (cf. Council's reluctance to formalise trialogue rules). Hence the contention among participants seems to be that the more informal a setting is, the more successful it is at reaching an agreement.

As we can see, several issues are intertwined here (i.e. rationality, legitimacy, efficiency, effectiveness, trust, etc.). The question, *more precisely*, is how and when it can be argued that closed settings have a legitimate place in a democratic decision-making process. As argued, the evaluative framework takes us some way towards providing the tools to answer these questions, but more research is needed in order to specify and disentangle the conditions under which closed and more informal settings can potentially be justified democratically. On a general level, Warren (2007: 286) argues that it is the *purpose* or *goal* of the setting that should decide the level of openness. Closed settings could be legitimate in situations where actors search for common ground and where a shielded setting is a means to reaching goals that can otherwise not be achieved. However, the main message conveyed in this book is that when the purpose of a setting is the adoption of public policy, secretly reached agreements must at some point be tested and justified in a publicly accessible manner. In other words, even if more shielded and informal settings can have a legitimate place in public decision-making, the balancing act between the open and formal versus the closed and informal should not tilt towards the latter. Privatising public policy-making is never compatible with democratic government.

Notes

1 Introduction

1 The procedure now includes policy areas such as fisheries, agriculture, structural funds, the entire budget and the former third pillar policy areas within security, freedom and justice.
2 In the following, I use the abbreviation OLP and co-decision-*making* interchangeably to refer to the procedure as it stands now under the Lisbon provisions. When I use co-decision *procedure*, this refers to the procedure prior to Lisbon. It should, however, be noted that the Lisbon Treaty did not introduce major changes to the procedural set-up (cf. Chapter 4) and that the name co-decision is still widely used both by researchers and practitioners.
3 See for instance, Crombez (1997, 2000); Häge and Kaeding (2007); Høyland (2006); König *et al.* (2007); Kreppel (2002); Napel and Widgrén (2006); Tsebelis (2002); Tsebelis and Garrett (1996, 1997a and b, 2000, 2001); Tsebelis *et al.* (2001); Kasack (2004); Kaeding (2005).
4 See e.g. the contributions in the 2011 special issue *West European Politics*, Vol. 34, No. 1.
5 Burns (2006); Corbett *et al.* (2007); Rasmussen (2005); Rasmussen and Shackleton (2005); Settembri and Neuhold (2008); Shackleton (2000); Shackleton and Raunio (2003). Some also look at the Commission's inter-institutional position under co-decision, see Rasmussen (2003); Burns (2004).
6 See e.g. Farrell and Héritier (2004); Häge and Kaeding (2007); Shackleton and Raunio (2003); Naurin and Rasmussen (2011); Héritier and Moury (2011).
7 The OLP is, however, part of the overall democratic evaluation of the EU system in Menéndez (2009), but is not the main topic of his assessment.
8 See, ESRC Project on the Informal Politics of Co-Decision: www.ucl.ac.uk/spp/research/esrc-project.
9 One exception is Bovens *et al.* (2010) where the concept of accountability is discussed and followed up by empirical testing. Accountability, however, only constitutes a part of the principle of democracy.
10 The discussion between liberals and communitarians is a common rotation point for these debates. See e.g. Rawls (1971, 1994); Habermas (1992, 1996, 2001); Young (1990); Walzer (1983, 1992, 1994); Miller (1995, 2000); MacIntyre (1981, 1988); Barry (1995).
11 See e.g. Weiler, 1995a and 1995b; Abromeit (1998); Eriksen and Fossum (2000, 2002); Eriksen (2009); Føllesdal and Hix (2006); Kohler-Koch and Rittberger (2007); Lord (2004, 2007b); Menéndez (2009); Cohen and Sabel (1997); Bohman (2005); Gerstenberg (1997), Miller (2009).
12 On the 'no-demos-thesis', see Weiler (1995a and b); Stie (2002).
13 For a discussion on different meanings with regard to the concept of accountability, see Bovens *et al.* (2010).

14 One example here could be the claim that EP elections are second-order elections, merely appendices to national ones in the sense that the outcomes of EP elections do tell us something about what the voters think of the performance of national politicians and are not a response to the performance of MEPs.
15 This was one of the main research questions of the now completed collaborative research project RECON (Reconstituting democracy in Europe) (http://reconproject.eu/). See also Curtin, 2009: 5–6.
16 Strong versus general publics (see Eriksen and Fossum, 2002) originally stem from Nancy Fraser (1992) who termed them strong versus weak publics.
17 Cohen and Sabel (2003); Eriksen (2003); Eriksen and Fossum, (2000); Fossum and Menéndez (2005); Joerges and Neyer (1997); Joerges (2006); Magnette, (2004); Rhinard (2002).
18 See for instance many of the contributions in Closa and Fossum (2004).
19 Cf. Habermas (1981a and b).
20 Here it could/would arguably be logical to label the potential deliberation taking place in expert and/or regulatory committees for epistemic deliberation. However, even if it is common to label the networks of experts working in the many EU regulatory committees for epistemic communities, the term 'epistemic deliberation' may cause confusion when discussed from the point of deliberative democracy. The problem is that the philosophical discussion on the epistemic dimension of deliberative democracy is (still) unresolved and roughly concerns the debate between correctness theories versus purely procedural accounts. The disagreement pertains to whether it is the process of deliberation in and of itself (i.e. the procedural account) that constitutes the quality of the outcome or whether it must also rely on procedure-independent principles (such as equality, liberty etc.) for setting up a fair process in the first place. If the latter is the case, the question becomes whether the whole process of deliberating is redundant for the overall achievement of democratically legitimate outcomes (See, Estlund, 1997). For these reasons, then, I find it more prudent and precise to label this type of deliberation elite deliberation instead of epistemic deliberation.
21 These issues are more closely presented in the second chapter, which contains a more thorough presentation of deliberative democracy.
22 Habermas (1996; 2001); Gutmann and Thompson (2002, 2004); Cohen (1999); Chambers (1996, 2003, 2004a and 2004b); Mansbridge (1992, 2003); Elster (1998a, b and c); Benhabib (1996).
23 Eriksen and Fossum (2000); Eriksen (2009); Closa and Fossum (2004).
24 Fishkin (1999); Steiner *et al.* (2004); Rosenberg (2007); Warren and Pearse (2008); *Acta Politica*, 2005, Vol. 40, No. 2 and 3.
25 That is:

> Deliberative democracy is focused on the circumstances in which a group must make a decision to which all members are bound whether they agree with it or not. Although even political deliberation can have various purposes, its essential aim is to reach a binding decision. From the perspective of deliberative democracy, other purposes – such as learning about issues, gaining a sense of efficacy, or developing a better understanding of opposing views – should be regarded as instrumental to this aim.
>
> (Thompson, 2008: 502–3)

26 Cf. the right to justification, Forst (2001); Chambers (2003: 308); Bovens (2007).
27 See also Ulbert and Risse (2005) who argue the same and share many of the same points. They are, however, not primarily concerned with democracy. Hence they do also not make a distinction between when arguing can generate democratic outcomes versus when arguing is merely generating more consensual outcomes.

2 What is democratic legitimacy?

1 For survey articles, see for instance Bohman (1998); Chambers (2003); Thompson (2008). There are also quite a few anthologies covering different theoretical positions, e.g. Bohman and Rehg (1997); Elster, (1998a, b and c), as well as some contributions on empirical approaches to deliberative democracy, see Steiner *et al.* (2004); *Acta Politica*, Vol. 40, No. 2 and 3, 2005; Rosenberg (2007); Warren and Pearse (2008).

2 I rely on an understanding of deliberative democracy which is not simply procedural, but which also acknowledges that the procedure itself is grounded on a certain minimum normative foundation. This foundation amounts to the principle of equal worth of persons, i.e. that for a decision-making procedure to be legitimate it must honour the principle that individuals are equals in the sense that they all have a right to justification when laws and regulations limit their actions. In other words, 'Procedural-independent standards are needed for securing a fair process' (Eriksen, 2009: 46). See Forst (2001) on the right to justification. For a discussion on procedure versus substance, see Gutmann and Thompson (2002); Bohman (1998).

3 However, even perfectly performed processes will not always end in consensus as there may be good reasons for people to stick to their original arguments and not change position. Hence the 'evidence' of whether a decision-making process has met the criteria of deliberative democracy cannot be the change of position alone (even if it can be an important and decisive indicator).

4 The problem is that:

> there is no rule for aggregating individual preferences that is obviously fair and rational and thus superior to other possible rules; and that virtually every rule is subject to strategic manipulation, so that even if it would produce a plausible outcome for a given set of preferences if everyone voted sincerely, the actual outcome is liable to be distorted by strategic voting.
>
> (Miller, 2000: 14).

5 Cf. also Mansbridge *et al.* (2010).

6

> A legitimate political order is one that could be justified to all those living under its laws. Thus, accountability is primarily understood in terms of 'giving account' of something, that is, publicly articulating, explaining, and most importantly justifying public policy. Consent (and, of course, voting) does no disappear. Rather, it is given a more complex and richer interpretation in the deliberative model than in the aggregative model. Although theorists of deliberative democracy vary as to how critical they are of existing representative institutions, deliberative democracy is usually not thought of as an alternative to representative democracy. It is rather an expansion of representative democracy.
>
> (Chambers, 2003: 308).

7 Cf. Schumpeter (1942); Downs (1957); Riker (1982).

8 See also Landwehr (2010) for a similar argument

9

> The reciprocal deliberations are *free* in the way that the participants are bound only by the results and premises of their own deliberation. The institutions are not tyrannical and do not impose norms and conditions on the citizens that they themselves do not approve of. The deliberations are *rational* in the way that the parties justify their standpoints and their proposals. It is a justificatory burden in relation to unequal treatment and it is the power of the arguments that governs collective decision-making. The parties are *equal*, i.e. all speech-competent individuals can participate and promote their views on equal terms. Differences in relation to power and resources should either be compensated for or eliminated. This can be done, for example, by means of appointing advocates for those who are in a less

> favourable position. The deliberation take place in *public*, and the outcomes of the deliberations are legitimate only when they have been approved of in a free discourse with identical or at least mutually acceptable reasons.
>
> (Eriksen, 2007a: 95–6).

10 In addition, it may also be that there is no consensus even when a process has been conducted in an impeccable manner, as there may be situations where actors hold equally reasonable and acceptable, but irreconcilable, arguments (see Eriksen, 2007a: 106–7).

11 Since rational consensuses are unlikely to achieve under non-ideal conditions also deliberatively reached decisions become – to some extent – unstable. Deliberatively reached agreements hold as long as a better reason does not come along, and, presumably, only as long as there is not sufficient time to consider what those better reasons are. However, deliberatively reached agreement are nevertheless more stable than compromises which are likely to change if there is a shift in the relative power relations (I am grateful to Chris Lord for reminding me about this point).

12 That is, reasons that are 'intersubjectively justifiable', they are different, but 'become understandable and mutually acceptable' through deliberative processes (Eriksen, 2007a: 107, 108).

13 Bohman (2005) defends a deliberative, polyarchical federalism as a solution for the EU. Keohane *et al.* (2009) argue along similar lines to Cohen and Sabel, but the former do not foresee or seem to believe in democracy at the supranational level. Rather they defend participation in multilateral (i.e. international) organisations as a means to enhance democracy at the national level.

14 For a discussion of whether democracy presupposes a national identity, see Stie (2002). For a discussion of whether democracy presupposes a state apparatus, see Eriksen (2009: Chapter 9); and Gaus (2008).

15 Moreover, the discussion on what deliberative democracy demands in empirical terms has resulted in indeterminacy of the very concept of deliberation. Although widely used, 'deliberation' as a normative as well as an analytical concept is not a precise term and is applied differently in various accounts. Authors use: 'public deliberation', 'political deliberation', 'democratic deliberation' (Neblo, 2005), 'epistemic deliberation', simply 'deliberation' but also 'arguing' (Ulbert and Risse, 2005), 'discursive behavior' (Rosenberg, 2007) etc. in a variety of ways. The point is that there is not yet an intersubjectively shared definition of what we mean when we talk about deliberation, as there is when we talk about bargaining. I shall not go more into detail on this as the focus here is on the *preconditions* for reaching deliberative-democratic decisions.

16 See particularly pp. 1003, 1005.

17 See also their contribution in Bächtinger *et al.* (2005).

18 Their case studies are parliamentary discussions in Germany, Switzerland, the UK and the US.

19

> in global governance direct stakeholder access to deliberation in international organisations is extremely difficult to realise, so that we rely on CSOs [i.e. civil society organisations] to transport arguments from stakeholders global deliberation. It is therefore essential for a democratic procedure that CSOs will have institutionalised access to these deliberative settings. Only in this way can it be secured that stakeholder arguments can be voiced.
>
> (Nanz and Steffek, 2004: 13).

3 Approximating ideal principles to non-ideal conditions: the evaluative framework

1 Habermas (1998, 2001); Eriksen (2003, 2007a and b, 2009).

2 See also Fraser (1992).

3 Apart from the fifth criterion, the other criteria are borrowed from Eriksen and Skivenes (2000). However, Eriksen and Skivenes' study is about the influence and impact of deliberative politics in public bodies such as the Norwegian Biotechnology Advisory Board and child welfare system. As these organs are not public policy-making bodies or law-making institutions, I have consequently adjusted the criteria in order to facilitate an assessment of the OLP. In short, I have thoroughly substantiated and operationalised the criteria and also developed and specified empirical indicators to accompany them.

4 As Farrell and Héritier (2003a: 577) note: 'Empirical evidence suggests that the Treaty articles governing codecision and other legislative procedures provide only a beginning point for interactions between the EP and the Council.'

5 In this sense there is an overlap with the institutionalist terminology in that 'institutions' are (formal or informal) 'sets of rules' and practices wherein actors (individuals and collective organisations) and organisations (collective actors such as the EP, Commission and Council) are situated. As this book is not based on an institutionalist approach as such, I do not follow its terminology in the sense that I often refer to the EP, the Council, the Commission etc. as the 'institutions'. This is in accordance with how the EU itself refers to these bodies. Moreover, in the text, if not otherwise mentioned, I use the terms 'institutions', 'institutional actors', 'bodies' etc. interchangeably. For institutionalist approaches, see March and Olsen (1989, 1998); Olsen (2007); Bjurulf and Elgström (2004); Farrell and Héritier (2004).

6 For a discussion on the relationship between democratic citizenship and bureaucracy related to the EU, see Olsen (2007: Part II).

7 Cf. the difference between institutionalised deliberation settings and the unorganised, spontaneous deliberations taking place in the general public spheres.

8 The question of actors is further dealt with under the next criterion concerning inclusion.

9 Cf. knowledge relevant for ethical-political and moral discourses in Habermas (1998: 168).

10

> The flip-side of inclusion and argumentation is the need to discipline discourses with regard to spatio-temporal and substantive claims. It seems fair to assume that a high degree of inclusion only translates into successful deliberation if it is accompanied by provisions which structure the discourse and limit the access of premises. Legal procedures regulate which topics and questions may be raised, the use of time, the participants, the distribution of roles etc. Modern law is a medium for discourse because it is not only a decision-making system but also because it contains principles for the allocation of the duty to argue
>
> (Eriksen and Neyer, 2003: 12).

11 The logical consequence of the participatory model is that it makes the legitimacy of political decisions dependent upon the actual participation of all (or at least a significant majority of) citizens. This is possible neither practically nor normatively. Normatively, because citizens also have a right not to participate or engage in politics if they so desire. It is, of course, detrimental also from a discourse theoretical perspective if citizens do not take part in elections and public life in general as even representative democracy is defined and dependent upon a certain level of participation and engagement to be deemed democratically legitimate (Eriksen and Weigård, 2003: 148–9).

12

> What is important to its notion of public reasoning is not so much that everyone participates but more that there is a warranted presumption that public opinion is formed on the basis of adequate information and relevant reasons, and that those whose interests are involved have an equal and effective opportunity to make their own interests (and their reasons for them) known. As an ideal, inclusion allows for maximum expression of interests, opinions and perspectives relevant to the problems and issues for which a political body seeks solutions
>
> (Nanz and Steffek, 2005: 370).

13 See also Gutmann and Thompson (2004: 9–10) for a deliberative, but not discourse-theoretical position.

14 Contrary to proponents of direct deliberative polyarchy, the version of deliberative democracy defended here requires that there must be some kind of congruence between affected parties and decision-makers and that this is not feasible without arrangements ensuring popular elections of representatives to the most important decision-making bodies. In other words, the availability of and access to networks and a plethora of deliberative settings are important, but not sufficient.

15 See in comparison Settembri, 2005. I share many of the same indicators, but they are not always interpreted in the same way. Settembri lists physical access to the buildings of decision-making bodies as one important indicator. I find this indicator less relevant as long as there is a digital register of documents as well as verbatim records or video transmissions of meetings available online. Online access to documents and debates levels out the playing field so that those not situated in, or in close proximity to, Brussels/Strasbourg are not disadvantaged. Hence, in my opinion, physical access is, in principle, redundant from the perspective of ensuring a transparent decision-making procedure, unless perhaps for the media.

16 By 'particular content' we are still on a principled level. The point is not to prescribe in detail (for instance whether a polity should have a consensus instead of a majoritarian system) how the organisational set-up should look like, but to ensure that certain foundational principles (e.g. separation of powers etc.) are respected.

17 This indicator is adapted to the EU context in the sense that since the Commission is not an executive with a popular mandate generated from EP or national elections, it is important to underline that the Commission should not also have final decision-making powers. Hence the importance of separating the policy presenter from the decision-makers is underlined from the perspective that the Commission should not both be agenda-setter and legislator and not because the EP and the Council should not have the right to initiate legislation. That democratically elected legislators should have the right to initiate legislation is a commonly agreed principle even if it is not so often a fact (see Corbett *et al.*, 2007: 238–9).

18 See also Føllesdal and Hix (2006: 548–9) for a similar argument in the EU context. Here they contest Majone's claim that the EU is sufficiently legitimate because it only deals with regulatory and not redistributive politics. Majone's argument hinges not only on the EU's ability to produce Pareto-efficient outcomes, but also that these decisions are strictly limited to the domain of regulatory politics. The critical point for Føllesdal and Hix, then, is how Majone can prove that the EU is not also taking decisions that are redistributive and hence create winners and losers. Their argument is, among other things, that it is not sufficient to look merely at policy outcomes without taking into account how such decisions came about.

19 Or in the words of Gutmann and Thompson (2004: 37): 'Deliberative democracy … is not intended to be a description of current political reality. It is an aspirational ideal' (See also Thompson, 2008).

20 Cf. Protocol 2 TFEU/TEU, 'On the Application of the Principles of Subsidiarity and Proportionality' (European Union 2010 in Official EU documents).

21 At the time of finishing the manuscript of this book (May 2012), the first yellow card procedure has been filed by the national parliaments on the so-called Monti II law. Cf. http://euobserver.com/851/116405.
22 E.g. Corbett *et al.* (2007); Collins *et al.* (1998); Shackleton (2000); Shackleton and Raunio (2003); Rasmussen and Shackleton (2005); Cortes (2000).
23 E.g. Farrell and Héritier (2003a and 2003b, 2004, 2007); Héritier (2003); Foilleux *et al.* (2005); Burns (2004, 2006); Hagemann and De Clerck-Sachsse (2007); Rasmussen (2003, 2005); Rasmussen and Shackleton, (2005); Rhinard (2002); Settembri and Neuhold (2008); Lewis (1998, 2000); Bjurulf and Elgström (2004); Maurer (2003); Tsebelis *et al.* (2001).
24 Corbett *et al.* (2007); Hayes-Renshaw and Wallace (2006); Hix (2005); Westlake and Galloway (2004) Nugent (2001); Richardson (2006).
25 Eriksen and Fossum (2002); Eriksen (2005); De Leeuw (2003); Føllesdal and Hix (2006) and partly Bovens *et al.* (2010) where the topic is accountability.

4 Background: from co-decision to the ordinary legislative procedure

1 For an overview of the situation prior to the Lisbon Treaty, see Stie (2010a: Section 4.3).
2 Official Journal 2010/C 83/01, http://eur-lex.europa.eu/LexUriServ/LexUriServ.do?uri=OJ:C:2010:083:FULL:EN:PDF.
3 If the Commission does not think that it will gain the sufficient majority either in the EP, or particularly in the Council or (worst case scenario) in both, the proposal is usually never tabled. This is why the success rate of OLP dossiers is so high and hardly any cases fail. For numbers, see Hayes-Renshaw and Wallace (2006: 213).
4 European Commission (II).
5 Prior to the Lisbon Treaty, the EP's stance was named 'Opinion' not 'Position' as is the case now. We may say that the change of words *formally* puts the EP on a par with the Council even if, since the Amsterdam Treaty, the positions of the two institutions have in *practice* carried equal weight in the legislative process (see below).
6 A simple majority is a majority of the MEPs voting, see EP (2007b: 7).
7 Or actually, decisions are most often adopted by consensus. I come back to this issue in more detail below.
8 With the exceptions of where the Treaties require unanimity (see above) and when the Council is disagreeing with the Commission's comments on the EP's amendments.
9 Here no absolute majority is needed. Høyland and Hagemann (2010) argue that this gives the Council an advantage, as it is harder for the EP to introduce its own amendments than it is to adopt the Council's Common Position.
10 After the coming into force of the Lisbon Treaty, at least 378 out of the total 754 MEPs must vote in favour in order to achieve an absolute majority. After the 2014 EP elections, the number of MEPs will decrease to 751 and an absolute majority will correspondingly be 376 MEPs.
11 An agreement is reached if the Joint Text is supported by an absolute majority of the EP delegation and a qualified majority in the Council delegation (Rasmussen, 2005).
12 Earlier, the chair of Coreper I was replaced by the permanent representative only in a very few cases and mostly in the financial field (Shackleton, 2000: 334). After 2005, a handful of legal bases previously figuring under the third pillar were moved to the first pillar and hence fell under the former co-decision procedure. The Lisbon Treaty abolished the pillar system and now the whole former third pillar is subjected to the OLP. As it is Coreper II that deals with justice and home affair issues because the permanent representatives are now more often involved in the OLP.
13 See European Parliament, Council and Commission (1999).
14 See European Parliament, Council and Commission (2007).

15 Earlier in the document it is even stated that 'The institutions confirm that this practice *should* be extended to cover all stages of the co-decision procedure' (my emphasis).

16 Cases 138/79 and 139/79, *The Isoglucose Cases* [1980] ECR. 3333; 3393. The 'isoglucose' ruling of the ECJ from 1980 states that the EP must be consulted prior to the Council taking a decision. In the ruling, the ECJ annulled the law adopted by the Council on the grounds that the Parliament had not yet given its opinion on the proposal. The Court linked the ruling to the question of democratic legitimacy:

> the means which allows the Parliament to play an actual part in the legislative process of the Community. Such a power represents an essential factor in the institutional balance intended by the Treaty. Although limited, it reflects at Community level the fundamental democratic principle that the people should take part in the exercise of power through the intermediary of a representative assembly. Due to consultation of the Parliament in the cases provided for by the Treaty therefore constitutes an essential formality, disregard of which means that the measure concerned is void.
>
> (ECJ ruling quoted in Corbett *et al.*, 2005: 199)

Another ruling by the ECJ in 1995 modifies this right and obliges the EU institutions to loyal cooperation (ibid: 200).

17 After the coming into force of the Lisbon Treaty assent is now called the consent procedure.

18 In the Lisbon Treaty this procedure has been abolished.

19 The EP acquired budgetary powers in the 1970s.

20 To distinguish between co-decision before and after the changes made in Amsterdam, I use the term 'co-decision I' to indicate the regulations under the Maastricht regime, whereas 'co-decision II' signifies the rules from Amsterdam and onwards (the Nice and Lisbon Treaty did not alter the basic workings of the procedure, only added new policy areas to the co-decision (but national parliaments)). Note, however, that when I use simply 'co-decision' elsewhere in the text, this means the current rules (i.e. Amsterdam co-decision II).

21 First Article 189b, then after the renumbering Article 251 (of the Treaty establishing the European Communities) and finally under the current Lisbon Treat; Article 294 (Treaty on the Functioning of the European Union, TFEU).

22 The EP majority was 377 votes in favour of rejection. Consequently,

> in 1998, when the conciliation committee failed to reach an agreement on a directive on a Securities Committee, Council decided to anticipate the provisions of Amsterdam and not take advantage of its right to challenge the Parliament to overrule its position
>
> (Corbett *et al.*, 2007: 215–6).

For a more thorough presentation and analysis of this case, see Rittberger (2000).

23 If the EP declared an intention to reject the Council's Common Position, the Conciliation Committee could be convened right away. 'This "petit conciliation" has only been used twice (Engine power in 1994 (COD 1991 371) and European Capital of Culture in 1999 (COD 1997 0290))' (Rasmussen, 2005: 1030, footnote 4).

24 Conciliation trialogues were, however, known from the budgetary procedure, but it is first and foremost under co-decision that they have gained momentum, Hayes-Renshaw and Wallace, (2006: 216).

25

> It is no surprise that Ministers prefer to send civil servants rather than attend themselves. Meetings can be highly technical and/or focused on very specific issues. In some cases, discussion has lasted for an hour or more over the position of one word!
>
> (Garman and Hilditch, 1998: 280)

26 See EP (2004: 4).

27

> Interestingly, both institutions anticipated this reform in March 1998, when no agreement on the comitology issue could be found within the conciliation committee on draft directives 93/6/EEC and 93/22/EEC. Instead of reshuffling the draft according to the Maastricht Treaty provisions, the two institutions declared the failure of the procedure after conciliation.
>
> (Maurer, 2003: 230)

28 However, as under the Maastricht provisions (and also under Amsterdam) there are no time constraints during the first reading, i.e. prior to the adoption of the Council's Common Position. This is also the rule today.

29 See European Parliament, Council and Commission (1999).

30 See Council (2000).

31 Cf. Article I-34(1) and III-396 (European Union 2004 in Official EU documents).

32 The Treaty establishing the European Community (TEC) is renamed Treaty on the Functioning of the European Community (TFEU) and co-decision or the ordinary legislative procedure is now found in Article 294.

33 See Protocol 2 of the TFEU/TEU, 'On the Application of the Principles of Subsidiarity and Proportionality'.

34 There is also the yellow card procedure applicable in policy areas falling outside the OLP. Here at least one-third of available votes (currently 18) must be cast against a proposal before the institution that presented it must review it. For proposals on judicial cooperation in criminal matters and police cooperation, the threshold is one-quarter of votes (i.e. 14 at present). Following review, the proposing institution may maintain, amend or withdraw it. It must give reasons for its decision. Since 2005, COSAC has organised pilot subsidiarity checks, whereby committees from all national parliaments have been asked to assess selected proposals against the subsidiarity and proportionality principles and to report on their conclusions. In contrast, the 'orange card' mechanism applies only to the OLP and it involves a higher threshold and more stringent consequences than the yellow card procedure. For more see, http://ec.europa.eu/codecision/stepbystep/text/index_en.htm.

35 In unitary systems, the parliaments have two votes whereas in bicameral systems, the upper and lower (or regional) houses get one vote each.

36 For an overview of adopted acts until 2002, see Maurer (2003).

37 Cf. EP Conciliations and Codecision website (last visited 7 May 2012). www.europarl.europa.eu/code/about/statistics_en.htm.

38 For sceptics, see list in Maurer (2003: 239).

39 In terms of length of OLP dossiers, that is, number of words, Settembri (2007) found that the 2004 enlargement has also involved an increase. Interestingly, this is not the case with legislative acts adopted under the consultation procedure.

40 In comparison, the average length of consultation acts has decreased from around 510 to 310 days.

41 In this period, three procedures failed, i.e. 2 per cent of the total number of OLP dossiers (ibid). Under Amsterdam there were 403 totally concluded OLP acts.

42 In two cases (0.5 per cent) no agreement was reached and the procedures failed (ibid). Under Maastricht there were 165 totally completed OLP dossiers.

43 See also EP (2007b: 9).

44 Cf. the EP's Conciliations and Codecision website (last visited 7 May 2010): www.europarl.europa.eu/code/about/statistics_en.htm.

45 Cf. European Parliament, Council, Commission (1999, 2007).

5 A forum for the public? Are there democratic deliberative meeting places?

1 When I, in the following sections on the EP, refer to rule numbers these are all taken from the EP's rules of procedure (EP, 2012).
2 Rule 50 in the EP's rules of procedure allows for so-called 'procedure with associated committees' where there may also be more than one committee involved (see also rule 51 and Settembri and Neuhold, 2008).
3 The representativity of EP committees is also confirmed in McElroy (2006: 25).
4 See also rule 187 on substitutes. The reason behind multiple memberships is that some committees are more influential and important than others and it would therefore be hard to get MEPs to commit themselves solely to those committees that are deemed less important (Corbett *et al.*, 2007: 128). Earlier, a seat in the Foreign Affairs and/or Budgets Committees was deemed the most high-status, but over the years committees with more legislative clout have become increasingly popular (Judge and Earnshaw, 2003: 185–6). This is the case for influential committees such as the Environment Committee and Economic and Monetary Affairs Committee which both handle a considerable number of OLP dossiers (ibid: 185–6, 187–8).
5 See, EP (I).
6 See Corbett *et al.* (2007: 335).
7 The EP also keeps a Legislative Observatory (OEIL) which is an internet access point providing a procedure file on every legislative dossier received and dealt with in Parliament. I come back to the OEIL in Chapter 7 under the criterion of openness and transparency.
8 Committee meetings are also open to visitors, but space is limited and the access has become much stricter after 11 September 2001. Lobbyists are the most active committee meetings visitors, followed by the press and casual visitors, according to Corbett *et al.* (2007: 334–40).
9 Due to the enlargements in 2004 and 2007 the pressure on interpretation has been particularly heavy. Hence rule 147 provides exceptions to the translation regime in a transition period ending with the seventh parliamentary period in June 2014. This exception was supposed to end with the sixth parliamentary term, but was renewed and extended to the end of the seventh parliamentary period.
10 See Title II, Chapters 1–6 covering the provisions on committee involvement in the OLP and Title VII, Chapters 1–2 on committee work in general (EP rules of procedure). In addition, temporary committees can be set up to deal with specific issues (cf. rule number 184). These committees operate outside the standing committee framework. Sub-committees and working groups, on the other hand, can be established within the permanent committee structure. Neither of these types of committees will be assessed here. For more information see Corbett *et al.* (2007: Chapter 7).
11 During the fifth Parliament (1999–2004) the number was 17, whereas the 1995–1999 Parliament had 20.
12 That is, the Environment, Public Health and Food Safety Committee (ENVI) covered 20 per cent of the dossiers, the Legal Affairs Committee (JURI) 18.3 per cent and the Transport and Tourist Committee (TRAN) 11.4 per cent. Due to the inclusion of justice and home affairs policy areas in 2005, the Civil Liberties, Justice and Home Affairs Committee (LIBE) had the largest increase. In the fifth parliamentary term it was responsible for eight dossiers whereas in the sixth parliamentary term this number rose to 38 OLP procedures.
13 This seems also to be confirmed by the MEPs themselves in the 'MEP Survey 2000' where they were asked to put the following positions in a prioritised order according to preference; president of their party group, leader of their national party delegation, President of the Parliament and committee chair (rapporteurships were not included in the survey). 'The chair of a committee came out as the most attractive position,

mentioned by 62 respondents as their first choice, followed by EP President (46), leader of the national party contingent (30), and chair of the party group (27)' (Mamadouh and Raunio, 2003: 339–40).

14 The allocation of posts between the political groups is based on the d'Hondt's system of proportional representation reflecting the size of the various groups. For more see, Mamadouh and Raunio (2003: 340); Bowler and Farrell (1995: 237).

15

> If a national delegation within a Political Group has already provided a President, Vice-President or Quaestor of Parliament, or the chair of its Political Group, its chances of gaining a major committee chair will normally diminish since other delegations will want to get their turn. In other cases, a national delegation may benefit from the distribution, as in 2004 when a single delegation within the EPP, the German CDU, was able to claim the chair of the two largest committees.
>
> (Corbett *et al.*, 2007: 130)

16 Cf. rule 192.

17 See, Mamadouh and Raunio (2003: 342); Benedetto (2005: 71). Also substitutes may take on rapporteurships or contribute to the drafting of the EP's position – especially if they have expertise and experience in the relevant policy field (Judge and Earnshaw, 2003: 186).

18 In the first reading, other committees can also propose amendments to the proposal. They must send their opinion to the responsible committee, where the rapporteur is in charge of incorporating them into the draft opinion to be presented in plenary, see rule 57 EP's rules of procedure.

19 Yordanova (2011) finds that ALDE has become more powerful to the detriment of PES in obtaining rapporteurships in the sixth parliamentary term.

20 It should, however, be noted that both in the sixth and seventh parliamentary terms, there were/are only 30 non-attached MEPs.

21 This finding is also supported by Neuhold (2007: 11–12), who found very similar numbers when studying the situation both before and after the 2004 enlargement. In both the fifth and sixth parliamentary term 'On average, a majority constantly above 90 per cent endorses reports under co-decision: 93,3 per cent and 92, 5 per cent in EP5 [i.e. fifth parliamentary term] and EP6 [i.e. sixth parliamentary term], respectively (all readings taken together).'

22 See rules 61–3 for second reading treatment in committee.

23 Voting in committee is regulated in rule 195 (EP's rules of procedure):

> (1) Any Member may table amendments for consideration in committee. (2) A committee may validly vote when one-quarter of its members are actually present. However, if so requested by one-sixth of its members before voting begins, the vote shall be valid only if the majority of the component members of the committee have taken part in it. (3) Voting in committee shall be by show of hands, unless a quarter of the committee's members request a vote by roll call. In this case the vote shall be taken in accordance with Rule 167(2). (4) The Chair may take part in discussions and may vote, but without having a casting vote. (5) In the light of the amendments tabled, the committee may, instead of proceeding to a vote, request the rapporteur to submit a new draft taking account of as many of the amendments as possible. A new deadline shall then be set for amendments to this draft.

24 When the report is moved from committee to plenary treatment it is then very hard for other MEPs to orient themselves on what each committee member has voted (Settembri, 2005: 648, see also Settembri and Neuhold, 2008: 237). More on this in Chapter 7.

25 One more concrete indication is that the committee may convene hearings or that the

Commission, the Council or other relevant actors can be invited to attend (see rule 193, EP rules of procedure).

26 Whereas the Commission is more used to and comfortable with explaining and defending its position, the Council has been reluctant to answer questions in committee meetings. Most MEPs – being ambivalent about the increased use of trialogues – prefer and have encouraged the Council to come to Parliament to present and defend its position during open committee and plenary sessions. The Council representatives have, however, preferred to use committee meetings as listening posts, but Corbett *et al.* (2007: 144) report that this is, although very slowly, 'beginning to change, with chairs of Council working groups sometimes being mandated to give Council's position on issues of substance'.

27 EP (2007b: 6).

28 The responsible DG and the General Secretariat are always represented at EP plenary sessions (Nugent, 2001: 150).

29 See EP (I) for the various plenary documents available online.

30 I come back to consider this point more closely in Chapter 7.

31 See EP's rules of procedure, Title VI for the general provisions guiding plenary sessions and Title II, Chapters 1, 3–6 for the provisions on legislative procedures.

32 Moreover, this rule also ensures that 'Members and political groups shall have direct access to the European Parliament's internal computer system for the consultation of any non-confidential preparatory document (draft report, draft recommendation, draft opinion, working document, amendments tabled in committee).' (EP rules of procedure).

33 The general provisions on voting (also on individual amendments) and quorum are found in Chapter 5 of the EP's rules of procedure.

34 Cf. rule 66(2): only amendments that that seek

> (a) to restore wholly or partly the position adopted by the Parliament in its first reading; or (b) [seeks] to reach a compromise between the Council and Parliament; or (c) [seeks] to amend a part of the text of a common position which was not included in – or differs in content from – the proposals submitted in first reading and which does not amount a substantial change within the meaning of rule 59; or (d) [seeks] to take account of a new fact or legal situation which has arisen since the first reading. The President's discretion to declare an amendment admissible or inadmissible cannot be questioned.

35 The provisions on voting and calling of quorum are detailed in Chapter 5 of the rules of procedure.

36 For the full story of the TDI group and a thorough discussion of the position of non-attached MEPs, see Settembri (2004).

37 Rule 149 of the rules of procedure provides the guidelines for the allocations of speaking time between political groups and the non-attached members.

38 Chapter 6 lists the instances where it is possible to by-pass the agreed speaking list.

39 Cf. Holzinger's (2005) findings in the German Bundestag.

40 See also Holzinger's (2005) study on the German Bundestag where she argues that the procedural set-up of parliamentary plenary settings is not necessarily likely to induce deliberation as we usually think of it, i.e. transformatively where the opponent is not only convinced by the other's argument, but is potentially also ready to act upon it. The reason for this is that plenary settings are usually the final decision-making setting where parliamentarians present arguments that are already well known as they have been debated in the committees where the more substantial discussions take place. This seemingly has an impact on the type of deliberation that is likely to occur in the respective settings (in this regard, Holzinger makes an analytical distinction between arguing and deliberation which describes debate in plenary and committee settings respectively). Plenary sessions serve a different purpose than committees and

they can be seen as arenas where MEPs present to the audience what they stand for and what they have defended. This is a form of public accountability where the deliberative task is sequenced between the committees and the plenary sittings.

41 Cf. rule 68 of the EP's rules of procedure.

42 Or more precisely, the Conference of Presidents decides a quota for each political group reflecting the composition in the Parliament and then the political groups themselves decide on an individual basis who their representatives are going to be – usually those who are members of the EP committee responsible for the dossier and thus familiar with the issues (Rasmussen, 2005: 1023). According to Rasmussen (2005: 1031, footnote 12):

> In the fifth session of the Parliament, 75 per cent of the full conciliation delegates were either full or substitute members of the committee which had dealt with the file prior to conciliation. If we exclude the permanent members, i.e. the vice-presidents, the figure rises to 92 per cent.

43 This is a unit in the EP Secretariat which steps in when the conciliation process is launched. The conciliation secretariat does not have the same sectoral knowledge of the particular case (like the units in the Secretariat that assist the MEPs during the first and second reading), but is a 'general service with an overall interest in ensuring that the conciliation system functions in such way as to secure the best possible outcome for Parliament in the negotiating process' (Rasmussen and Shackleton, 2005: 12). See also: www.europarl.europa.eu/code/information/mission_en.pdf.

44 European Conservatives and Reformist Group.

45 The Vice Presidents were included in the informal talks in 1999:

> In order to improve effectiveness, fully monitor negotiations and ensure greater transparency, it was thought desirable to increase the number of persons taking part in direct negotiations with the Council outside actual meetings of the Conciliation Committee. This ensures greater continuity in relations between our Institution and the Council and strengthens Parliament's role in the procedure
>
> (EP, 2007a: 13)

One of the three Vice Presidents meets on all conciliation meetings for a period of 12 months. These representatives are selected by the political groups and must represent at least two different political groups.

46 Rasmussen (2008: 103) notes that after enlargement to 27 member states, the EPP and the former PES:

> are even more overrepresented [than in the fifth parliamentary term] because the extreme right and left groups … plus the non-attached have been excluded from conciliation work altogether. This means that the PPE and the PES groups have one seat more than they would have had if strict proportionality between the plenary and the delegation had applied, and all the other groups except the liberals are underrepresented.

47 Go to the heading 'ongoing conciliations', where the names of the MEPs are provided along with other information on each individual dossier, see EP (II).

48 The 'joint four-column working document' consists of four columns (or five columns given that the Commission's opinion on EP second reading amendments is known) containing the Council Common Position, the EP second reading amendments, the Council's position on the EP's second reading amendments and the EP delegation position.

49 For access, see, EP (I).

50 The purpose and tasks of the EP delegation are further described in EP (2007b).

51 See, www.europarl.europa.eu/code/about/activity_reports_en.htm, www.europarl.europa.eu/code/information/guide_en.pdf.

52 See, EP (2007b).
53 After the coming into force of the Lisbon Treaty, the number of Council configurations has been increased from nine to ten due to the splitting up of the General Affairs and External Relations Council into two separate configurations: (1) General Affairs; (2) Foreign Affairs; (3) Economic and Financial Affairs; (4) Justice and Home Affairs; (5) Employment, Social Policy, Health and Consumer Affairs; (6) Competitiveness (internal market, industry, research and space); (7) Transport, Telecommunications and Energy; (8) Agriculture and Fisheries; (9) Environment; (10) Education, Youth, Culture and Sport. The list of Council configurations is regulated in Annex I in the Council's rules of procedure. Whereas the General Affairs Council has a co-ordinating role when differences of interests between the other Councils conflict, the most prominent of the sectoral configurations are the Agriculture (and Fisheries)-, the ECOFIN- and the Competitiveness Councils. Although to varying degree, all configurations apart from the Foregin Affairs Council deal with OLP dossiers and although they all have distinct features of their own, I will not highlight them here (but see Hayes-Renshaw and Wallace, 2006: Chapter 2).
54 See Council (2006d).
55 See, Council (I).
56 See, Council (II).
57 See, Council (III).
58 See, Council (IV).
59 In general, Council minutes include the name and number of the dossier, the relevant documents, legal basis, the outcome, statements and result of voting (cf. Article 9 and 13 in the rules of procedure, cf. also Westlake and Galloway, 2004: 348).
60 These statements are optional and can thus be used strategically by the member states and therefore do not provide a systematic source to detect disagreement within the Council. Nevertheless, they can provide the public with an indication of where member states disagreed.
61 See, Council (V).
62 Cf. Annex II (particularly Articles 10 and 11, Council of the European Union, 2009) in the rules of procedure, the Council is obliged to keep an online register of documents where documents intended for publication in the Official Journal are publicly available.
63 Full verbatim records of ministerial meetings ended during the 1970s (Westlake and Galloway, 2004: 348).
64 Cf. Article 3(6) Council rules of procedure stating that:

> The items appearing in each part of the provisional agenda shall be divided into A items and B items. Items for which approval by the Council is possible without discussion shall be entered as A items, but this does not exclude the possibility of any member of the Council or of the Commission expressing an opinion at the time of the approval of these items and having statements included in the minutes.

65 The trialogues are, at best, semi-formalised. I come back to this in the following chapters.
66 Earlier, it was first and foremost the deputy permanent representatives from Coreper I who were involved in co-decision-making, but with the expansion of the number of policy areas falling under the OLP, Coreper II is now much more or equally involved in OLP dossiers.
67 Navarro (Spanish Deputy Permanent Representative in Coreper (I)), see EP (2002).
68 Article 294(11) TFEU; see also Rasmussen (2005); König *et al.* (2007: 291).
69 The EP offers access to conciliation documents ex post, see EP (I). Being a formal decision-making arena, the Conciliation Committee operates in all the official EU languages.
70 If successful, the EP, the Council and the Commission usually also host a joint press

conference to announce the result of the Conciliation Committee (cf. EP, 2007b: 17–8).

71 See EP (III), see also, EP (I).

72 An absolute majority is required for the EP delegation and a qualified majority is required for the Council delegation (Article 294(10)). However, Article 294 does not specify the type of majority required from the EP delegation for a Joint Text to pass, but 'After the failure of the takeover directive (...) the legal service of the EP clarified that the delegation needs not merely a simple but an absolute majority to adopt the joint text' (Rasmussen, 2008: 111, footnote 3).

73 Cf. the IATA case (C-344/04).

74 Cf. European Parliament, Council, Commission (2007).

75 Hence, a Conciliation Committee meeting thus usually deals with more than one case at each meeting.

6 Inclusion of affected and competent parties

1 However, the long and complex consultation process before the Commission formally tables a dossier provides the Commission with a good indication of the viability of a policy proposal. The 'success' rate of adopted dossiers indicates that the Commission does not forward legislation it does not believe will pass.

2 Apart from the Council meetings where internal matters are discussed (Hayes-Renshaw and Wallace, 2006: 34). I come back to the Commission's powers under the fourth criterion on the indicator on separation of powers.

3 Of the 85 per cent, 10–15 per cent is decided at the Coreper level, cf. Hayes-Renshaw and Wallace (2006: 79); Hix (2005: 77); Van Schendelen and Scully (2006: 9).

4 See also Table 3 in Annex III in Settembri (2007).

5 Even if the officials in the EP Secretariat also assist their MEPs quite closely, the type of involvement by the various departments in the EP Secretariat cannot be compared to that of the Council (or the Commission for that matter). See Neunreither (2002) for an introduction to how the various departments assist the MEPs.

6 The quote originates from the EP's 'Delegations to the Conciliation Committee: Activity Report 1 August 2001 to 31 July 2002', but is cited from Rasmussen and Shackleton (2005: 15) as the report is no longer available on the EP website.

7 See EP (2009b: 34–5).

8 As we have seen, Settembri (2007) even talks about 'giant coalitions' as he found that the OLP reports adopted in EP committees were generally supported by over 90 per cent of the MEPs.

9 This is not only due to the post-Maastricht lobbying boom and the fact that the sheer number of external actors participating and approaching the Commission has steadily risen to a level where it is practically impossible to meet them all with equal consideration and intensity (Broscheid and Coen, 2003). This 'overload' or 'oversupply' of lobbyists has led to a 'management' problem to which the Commission has responded by introducing a more institutionalised consultation system which seemingly reflects the difficulty in balancing the need for information with openness and equal access. Prior to the latest enlargements, Broscheid and Coen (2003: 167) estimated that 'The Commission bureaucracy has approximately 16,000 members, the size of a larger city administration.'

10 See Quittkat and Finke, 2008; see the contributions in Kohler-Koch and Maloney (2008). See also the Commission's consultation website (European Commission (III)).

11 The lack of grassroots anchoring is also indicated in a news article in *European Voice*, 21 April 2005.

12 Moreover, 'On a system-wide level, more participation may increase the overrepresentation of those who are already well represented, generating the paradox that

increasing citizen opportunities for participation may increase political inequality' (Warren, 2008: 56).

13 Gornitzka and Sverdrup (2008: 15) found that 'it is about fifty per cent chance that you will find only national officials seated around the table.... [Moreover,] expert groups composed only by societal actors or only by scientists are rarely found.' See also Larsson and Murk (2007); Egeberg *et al.* (2006).

14 Moreover, the Commission has 'strategically fostered interest group participation that would be consistent with and conducive to further integration' (Mahoney, 2004: 443). With regard to the Commission's funding regime, Quittkat and Finke (2008: 214) found that 'the European Commission mainly supports associations representing either the weak or excluded, whereas CSOs representing general interests are less often supported.'

15

> the Parliament is often criticised not only for its inability to understand the constraints of the legislative procedure but more seriously for being uninformed or subjectively informed about the issues involved. The sources of information for European parliamentarians are often challenged by actors operating in working groups who consider that they have better, i.e. 'more objective', information than parliamentarians do. The latter are often stigmatised for supposedly relying upon information from self-interested lobbies and private companies. The second point, partly as a consequence of the first, is that most working group members regret the time 'wasted' by the new procedures.
>
> (Fouilleux *et al.*, 2005: 617–8)

16 Cf. in the year 2010, 80 dossiers were adopted at first reading whereas 14 and zero were adopted in the second and third readings respectively (European Commission, 2011: 130).

17 See Farrell and Héritier (2003b: 24) and Bunyan (2007a and b).

18 Cf. Article 19, Council rules of procedure (Council of the EuropeanUnion, 2009) See list of other documents.

7 Openness and transparency

1 Prelex service (see European Commission (IV)), OEIL service (see EP (IV)), EUR-Lex service (see internet link in list of Official EU documents).

2 See, Council (VI).

3 Cf. ex Article 255 TEC (European Union, 2006).

4 See for instance, Davis, (1999); De Leeuw (2003); Driessen (2005); Dyrberg (2002); Heliskoski and Leino (2006); Öberg (1998, 1999); Curtin (2000, 2001, 2003); Peers (2002, 2006); Settembri (2005). The Regulation is currently under re-evaluation, see Commission (2008b), EP (2009c) as well the procedure file on the EP's Legislative Observatory: www.europarl.europa.eu/sides/getDoc.do?pubRef=-//EP//TEXT+TA+P6-TA-2009–0114+0+DOC+XML+V0//EN&language=EN. See also Statewatch's Observatory on the Regulation on access to documents: 2008-ongoing (for internet link, see list of Official EU documents). There has also been quite a lot of input from the public sphere on the issue EU transparency. A selection of media articles from EurActive, European Voice, Statewatch and Access Info can be found in the list of Official EU documents.

5 I come back to this under criterion 4 and the indicator on intelligibility of procedure.

6 See Access Info (2012); Statewatch Observatory on the Regulation on access to documents: 2008-ongoing; see also Council (2012).

7 For Commission register, see European Commission (V); for EP register, see EP (V); for Council register, see Council (VI).

8 The easiest way to get an overview of all OLP dossiers either completed or still under

consideration and with a direct link to the relevant Prelex site, is to access the Commission's co-decision homepage (see European Commission (VI)).

9 See, Council (VII).

10 See European Commission (VII). 'C' documents relate to official instruments for which the Commission has sole responsibility. Some are transmitted to the Council or Parliament for information.

11 See, EUR-Lex (for internet link, see 'Other EU/inter-institutional internet sites' in list of Official EU documents).

12 See EP (2011: 3).

13 It could possibly be argued that access to comitology documents is indirectly important in the sense that after an OLP act is officially adopted it moves to the implementation phase where it is dealt with in the many comitology committees. However, while this could be important, the question here is whether the documents produced during the legislative process as prescribed in Article 294 are accessible to the public.

14 For a brief historical presentation of the transparency developments of the Council, see Bauer (2004: 367–72).

15 According to Corbett *et al.* (2007: 182), due to the introduction of the video transmission service, full records of speeches are no longer (since September 2007) available.

16 See EP (III).

17 See, EP (VI).

18 As an example, the access point for the meeting documents of the Committee on Economic and Monetary Affairs, see EP (VII).

19 See, EP (VIII).

20 See, EP (IX).

21 Cf. Article 7, Council rules of procedure.

22 Bauer (2004: 368) notes, however that:

> One qualification … to this apparent veil of secrecy is the fact that Brussels journalists and lobbyists could always get a blow by blow account of what had gone on in Council meetings through press conferences and contacts with officials and civil servants. Indeed, there are few more leaky organisations than the European institutions. This point does not change the fact that for the general public it is still hard to keep track on preparatory process in the Council.

23 But note also that:

> Items adopted in committee by an overwhelming majority (fewer than one-tenth against) are taken in plenary without debate (unless requested by a Group of 40 members) and are subject to a single vote without amendments (unless one-tenth of the members – individually or by Group – requests that is be open to amendment).
>
> (Corbett *et al.*, 2007: 172)

24 Secret ballots are rare and have not been used in OLP processes. They are, however, obligatory for the elections of the Commission President and Ombudsman, as well as the President, Vice-Presidents and Queastors of the EP (Corbett *et al.*, 2007: 176).

25 See rule 167 EP's rules of procedure.

26 Cf. Neuhold (2007: 7) referring to Hix, Noury and Roland, reports that one-third of the votes cast are roll-call votes. It should also be noted that since roll-calls are optional, they are often used strategically for instance to embarrass and highlight internal divergence in other political groups and/or to signal to the Council the size of the majority supporting the EP's position. For this reason, it should be remembered that the results from studies merely based on roll calls must be treated with caution as MEPs may vote differently when they know that their individual vote will be publicly recorded. For studies on roll calls, see, Carubba *et al.* (2004); Hix *et al.* (2005, 2006).

27 For an overview of the historical development of Council voting rules, see Hayes-Renshaw and Wallace (2006: Chapter 10).
28 See, Council (IV).
29 This service has been available since 1997 (Hayes-Renshaw and Wallace, 2006: 278).
30 What is more, voting results are categorised according to Council configuration and not (for instance) according to policy area, decision-making procedure or voting rule (unanimity or QMV) (Mattila, 2004: 35, 48). For numbers (until 2004) according to member states as well as policy area, see Tables 10.3 and 10.4 in Hayes-Renshaw and Wallace (2006: 283–4).
31 See Statewatch (2001a). Moreover, right after an agreement was reached, Danish MEP Jens Peter Bonde approached the Council in order to obtain the minutes of the negotiations and concurrent drafts of the Regulation. The Council rejected Bonde's request and could not even provide him with the number and dates of informal meetings, see *European Voice*, 10 May 2001.
32 In addition, the 2007 Joint Declaration states that the working documents shall be publicly available from the institutions' registers after a dossier is concluded, (see, European Parliament, Council and Commission, (2007), rule 37). It is also possible to follow the work in the conciliation committee on the internet as the EP provides a separate link with information about what conciliation under the OLP is, it provides a calendar of upcoming conciliation meetings, short summaries of ongoing and concluded procedures as well as meetings documents (see, EP (III)). There are no such calendars of trialogue meetings.
33 This is also a shortcoming in the Regulation 1049/2001 itself in that it does not commit the institutions to establish such a register (see, European Parliament and Council of the European Union, 2001). As mentioned above, the Regulation is under re-consideration and already in 2006, the EP urged the Commission to table a proposal to improve Regulation 1049/2001 (see EP, 2006). Among the suggestions is a common access point for preparatory documents, connecting the institutions' registers in a better way, agreement on common rules for archiving documents and making it easier to follow decision-making processes. The EP has upheld its position of 'a single EU portal to ensure access to documents' in its first reading report – see suggested amendment to Article 12(4), in EP (2009). Steve Peers writing for Statewatch does not believe that the Commission's suggestion represents an improvement and in certain areas is rather to the contrary (see several of his articles at Statewatch's Observatory on the Regulation on access to documents: 2008–2009, for internet link, see list of other documents).

8 Neutralisation of asymmetrical power relations

1 Cf. e.g. the Commission's overview of the legal bases in their online OLP guide: http://ec.europa.eu/codecision/docs/legal_bases_en.pdf or under the heading 'Background documents' on the EP's website 'Conciliation and Codecision': www.europarl.europa.eu/code/information/legal_bases_en.pdf.
2 See Commission (VIII); Council (VIII); EP (2009a). In addition, the EP's Conciliation and Codecision Secretariat has a link on the EP's website for information about conciliation processes, see, EP (X).
3 See European Parliament, Council and Commission (1999, 2007).
4 See also Nugent (2001: 236–7); Chalmers *et al.* (2006: 96).
5 It should be noted, however, that the European Council has a unique position in the Union and has the capacity to initiate and move politics in a desired direction due to its membership consisting of the heads of state and government. Because of the political standing of the European Council, the Commission really has no choice but to follow its instructions (Nugent, 2001: 239). Moreover, with the Lisbon Treaty the European Council has obtained a formal status as an EU institution (European Union

2010: Article 13(1) TEU) and has now also a formal right to 'define the strategic guidelines for legislative and operational planning within the area of freedom, security and justice' (European Union 2010: Article 68 TFEU).

6 See also Nugent (2001: 261); Burns (2004); Rasmussen (2003); Mattila (2004: 32).

7 It is also indicated that the Commission is more influential when the OLP dossiers are of a more technical nature:

> Moreover, the Socrates cases cannot be regarded as particularly easy for the Commission to influence compared to more technical policy areas where its high level of expertise in relation to the Parliament and the Council is typically a greater strategic advantage.
>
> (Rasmussen, 2003: 2; see also Burns, 2004)

8 EP (2007b: 7).

9 See comment by Navarro who was Spanish Deputy Permanent Representative in Coreper (I), see EP (2002).

10 In her study, Burns (2004: 2) finds that in comparison to other decision-making procedures, the:

> Commission's formal and informal influence has declined under this [i.e. the OLP] legislative procedure. However,... many of the theoretical claims made about the Commission's relative influence overstate its weakness as they overlook the institution's agenda-setting and gate-keeping roles.

11 More information on how the Court can exert this right is provided under the next indicator.

12 E.g. an assessment of how the wider EU context affects the power relations between the institutional actors, not least concerning the Council–EP relationship as the former is by far the more powerful actor seen from a more general Union perspective. In addition, the Commission is not properly accountable to the EP and an appraisal of the institutional set-up pointing at the lack of a government–opposition dynamic between the Commission and the EP could also have been included (see Schmitt, 2005: 699; Føllesdal and Hix, 2006: 536). A more thorough analysis should also have included the impact of the Comitology system (for analyses of the Comitology system prior to Lisbon, see e.g. Joerges and Neyer, 1997; Joerges and Vos, 1999; Vos, 2009; Menéndez, 2009; Curtin, 2007a. For an overview of the new Comitology system, see Hardacre and Kaeding, 2011). For a discussion on the accountability of the EU system in general, see Lord (2004: Chapters 6–8).

13 Cf. the preliminary ruling procedure, Article 267 TFEU.

14 The possibility for individual citizens to file a complaint at the European level – the consequence of the principle/doctrine of direct effect – is one of the important features distinguishing the EU from intergovernmental organisations. Cf. also Hix (2005: 122–3).

15

> The Court of Justice of the European Union shall have jurisdiction to give preliminary rulings concerning: (a) the interpretation of the Treaties; (b) the validity and interpretation of acts of the institutions, bodies, offices or agencies of the Union; Where such a question is raised before any court or tribunal of a Member State, that court or tribunal may, if it considers that a decision on the question is necessary to enable it to give judgment, request the Court to give a ruling thereon. Where any such question is raised in a case pending before a court or tribunal of a Member State against whose decisions there is no judicial remedy under national law, that court or tribunal shall bring the matter before the Court. If such a question is raised in a case pending before a court or tribunal of a Member State with regard to a person in custody, the Court of Justice of the European Union shall act with the minimum of delay.
>
> (European Union, 2010: Article 267 (TFEU))

16 Cf. the principles of direct effect and supremacy which constitute the immediate legal foundation for the importance of the preliminary reference procedure (Hix, 2005: 128).
17 Cf. Case 25/62 *Plaumann & Co* v *Commission* [1963] ECR 95. For a more nuanced discussion on the application of Article 263, see Chapter 10 in Chalmers *et al.* (2006).
18 Title borrowed from Magnette (2003).
19 The Ombudsman is appointed (by majority vote) by the EP for a five-year term, which coincides with the Parliament's own term. The EO is under the direct authority of the Parliament, but should conduct his work independently of instructions or pressure from others, including the EP (but of course within the parameters of his legally defined duties). The ECJ can dismiss the EO upon request from the Parliament if the EO does not fulfil his duties or in situations of misconduct. The first Ombudsman took office in 1995.
20 See also the Statute of the Ombudsman office laid down in an EP Decision (cf. European Parliament, 1994).
21 See, European Ombudsman (I).
22 The first Ombudsman was the former Finnish Ombudsman Jacob Söderman. He retired in 2003 and the current Ombudsman, Nikiforos Diamandouros, then took over. Diamandouros has prior Ombudsman experience, as he held the Greek equivalent position up until becoming the European Ombudsman.
23 See, European Ombudsman (II).
24 Consequently, decisions made by national, regional or local authorities (including national courts and ombudsmen) as well as actions by private firms and individuals fall outside the scope of his jurisdiction.
25 The Ombudsman's Special Report provides a good summary to the whole process (2005b). See also the EO's Draft Recommendation to the Council (2005a).
26 The Council also held that the complaint was beyond the jurisdiction of the Ombudsman.
27 Cf. Council (2006a).

9 Decision-making capacity

1 Cf. Chapter 4. 'Co-decision I' defines the co-decision rules under the Maastricht Treaty and 'Co-decision II' signifies the revised co-decision rules after the Amsterdam treaty reform.
2 These were:

> the directive on the patenting of biotechnological inventions in 1995 where the plenary of the Parliament declined to ratify the results of the conciliation negotiations, and the Securities Committee directive in 1998 where the Council took a decision not to reintroduce the common position in anticipation of the Amsterdam provisions.
>
> (Shackleton, 2000: 237)

3 Contrary to the majority of commentators, Moravcsik and Nicolaïdis provide an intergovernmental view in holding that the:

> member states' decision to increase the powers of the EP was largely a consequence of left-wing victories in France and Germany. Thus, they imply, the EP's efforts to remove from the path of play the Council's right to reintroduce its common position was irrelevant to the later decision by the member states at Amsterdam formally to remove it.
>
> (Farrell and Héritier, 2003a: 589).

4 Even if it is not stated explicitly in the latest Joint Declaration on the OLP that trialogues are necessary for reaching agreement, their importance is not to be

misunderstood as, for instance, they 'must continue to be encouraged' (European Parliament, Council and Commission, 2007).

5 Whereas Article 294 is still silent in this regard, the Joint Declaration states that:

> Trilogues shall take place *throughout* the conciliation procedure with the aim of resolving outstanding issues and preparing the ground for an agreement to be reached in the Conciliation Committee. The results of the trilogues shall be discussed and possibly approved at the meetings of the respective institutions.
>
> (European Parliament, Council, Commission, 2007, rule 25, my emphasis)

6 See e.g. Chalmers *et al.* (2006: 151); Collins *et al.* (1998: 8); Rasmussen (2005: 1025); Bostock (2002: 222).

7 See, European Parliament, Council, Commission (2007: rule 30).

8 However, in the IATA case (Case C-344/04 and ELFAA 2006: ECR 1-403), the ECJ ruled that:

> The wording of Article 251 EC [i.e. now Article 294, TFEU] does not ... itself include any restriction as to the content of the measures chosen that enable agreement to be reached on a joint text. In using the term 'conciliation', the authors of the Treaty intended to make the procedure adopted effective and to confer a wide discretion on the Conciliation Committee. In adopting such a method for resolving disagreements, their very aim was that the points of view of the Parliament and the Council should be reconciled on the basis of examination of all the aspects of the disagreement.
>
> (paragraphs 56–7).

9 Cf. EP Conciliations and Codecision website (last visited 7 May 2012), www.europarl.europa.eu/code/about/statistics_en.htm.

10 Whereas the Council has never rejected an OLP dossier after conciliation, the EP has voted down two dossiers (the Takeovers Directive in 2001 and the Port Services Directive in 2003) (EP, 2009b: 24).

11 European Parliament, Council and Commission (2007: rules 2 and 8).

12 Ibid, rule 11.

13 As mentioned in Chapter 6, in 2004 the EP published guidelines for best practice on how MEPs should behave when participating in first and second reading trialogues with the Council (see EP, 2007b: 10). These guidelines were updated in 2008 and are now titled 'EP Code of Conduct for Negotiating Codecision Files' (see EP, 2009b: 42–4).

14 On a more general note, the Code states that:

> As a general rule, Parliament shall make use of all possibilities offered at all stages of the codecision procedure. The decision to seek to achieve an agreement early in the legislative process shall be a case-by-case decision, taking account of the distinctive feature of each individual file. It shall be politically justified in terms of, for example, political priorities; the uncontroversial or 'technical' nature of the proposal; an urgent situation and/or the attitude of a given Presidency to a specific file.
>
> (EP, 2009b: 43).

15 Rasmussen and Shackleton here commented upon the Guidelines from 2004, but the same could be said about the updated Code of Conduct from 2008, as this is also not legally binding upon the MEPs.

16 For instance Regulation 562/2006 establishing a Community Code on the rules governing the movement of persons across borders (Schengen Borders Code) and Regulation 1987/2006 on the establishment, operation and use of the second generation Schengen Information System (SIS II) (see, European Parliament and Council of the European Union, 2006a and b).

17 MEPs with good personal links to Council officials may exploit such channels for information. Høyland (2006: 31) has also found that there is seemingly a link between the degree of influence of EP rapporteurs depending on whether they come from the same political party as the ministers in the Council: 'Compared with rapporteurs from parties not represented in the Council, EP rapporteurs from parties represented in the Council may incur lower costs in coordinating their proposals with the informed actors in the Council.'

18 Farrell and Héritier (2003b: 28) mention the Takeover Directive as an example of this phenomenon.

10 OLP – democratic decision-making or technocracy in disguise?

1 This even applies to in the EU's foreign and security policy. See, Lewis (1998, 2000); Curtin (2007b); Smith (2003); Duke and Vanhoonacker (2006); Sjursen (2011); Cameron (2007).

2 There are different ways of conceptualising the early warning system. Cooper (2012) hypothesises whether it could trigger the national parliaments to form and function as a 'virtual third chamber'. Kiiver (2011: 101–2; see also Buzogány and Stuchlík, 2011), on his part, suggests that the system could be compared to a French-style Conseil d'Etat, which functions as:

> a consultative body to which the government submits its bills before it submits them to parliament.... In other words, the EWS is designed as a relatively narrow advice procedure, similar to the advice procedure applicable to councils of states regarding bills, and empirical analysis reveals that several national parliaments in facts use the EWS in such a manner.

3 For instance, they concern cases on mandatory data retention, access to documents, privacy in telecommunications etc. (see Bunyan, 2007b).

4 Having said this, it should, as outlined in Chapter 3, be remembered that in focusing on procedural qualities the conclusions that can be drawn tell us something about the *preconditions* and thus the *likelihood* of *democratic* decision-making taking place. In other words, the conclusions about these general procedural characteristics of co-decision-making cannot automatically be transferred to how individual OLP dossiers have been decided, but should be assessed on their own merits.

5 Steiner *et al.* (2004) found a mixed result when assessing the effect of publicity on open versus closed settings.

6 In relation to this, Thompson (2008: 510) also reminds us that:

> the empirical conditions of publicity must be distinguished from the normative requirement that deliberation be conducted in terms of public reason. This requirement, the scope of which is controversial among theorists, is a conceptual criterion of deliberation, as noted above in the discussion on public-spiritedness. The two should be kept distinct because the empirical condition of publicity may affect the extent to which the conceptual requirement can be satisfied. It is important to know whether this hypothesis or its opposite is valid: The more public the discussion, the more likely the participants are to use public reason, and the more likely the discussion is to be deliberative.

7 The need for confidential meetings in the Council and European Council are justified by the same reason.

Bibliography

Abromeit, H. (1998) *Democracy in Europe Legitimising Politics in Non-State Polity*, New York: Berghahn Books.

Ackerman, B. and Fishkin, J. (2004) *Deliberation Day*, New Haven: Yale University Press.

Alexy, R. (1989) *A Theory of Legal Argumentation: the Theory of Rational Discourse as Theory of Legal Justification*, Trans. R. Adler and N. MacCormick, Oxford: Clarendon Press.

Andersen, S. (1997) *Case-studier og Generaliseringer. Forskningsstrategi og Design*, Bergen: Fagbokforlaget.

Andeweg, R. (2007) 'A Comment on Auel, Benz and Maurer' in Kohler-Koch, B. and Rittberger, B. (eds) *Debating the Democratic Legitimacy of the European Union*, US: Rowman and Littlefield, pp. 102–9.

Bächtinger, A. and Steiner, J. (2005) 'Introduction', *Acta Politica*, 40(2): 153–68.

Bächtinger, A., Spörndli, M., Steenbergen, M.R. and Steiner, J. (2005) 'The Deliberative Dimensions of Legislatures', *Acta Politica*, 40(2): 225–38.

Barry, B. (1995) *Justice as Impartiality*, Oxford: Clarendon Press.

Bauer, M. (2004) 'Chapter 20: Transparency in the Council' in Westlake, M. and Galloway, D. (eds) with Digneffe, T., *The Council of the European Union*, 3rd edition, London: John Harper Publishing, pp. 367–88.

Benedetto, G. (2005) 'Rapporteurs as Legislative Entrepreneurs: the Dynamics of the Codecision Procedure in Europe's Parliament', *Journal of European Public Policy*, 12(1): 67–88.

Benhabib, S. (1996) *Democracy and Difference: Contesting the Boundaries of the Political*, Princeton, NJ: Princeton University Press.

Beyers, J. and Dierickk, G. (1998) 'The Working Groups of the Council of the European Union: Supranational or Intergovernmental Negotiations?', *Journal of Common Market Studies*, 36(3): 289–317.

Bishop, M. (2004) 'The Council's Legal Service' in Westlake, M. and Galloway, D. (eds) with Digneffe, T., *The Council of the European Union*, 3rd edition, London: John Harper Publishing, pp. 357–66.

Bjurulf, B. and Elgström, O. (2004) 'Negotiating Transparency: The Role of Institutions', *Journal of Common Market Studies*, 42(2): 249–69.

Blichner, L.C. and Molander, A. (2008) 'Mapping Juridification', *European Law Journal*, 14(1): 36–54.

Bohman, J. (1998) 'Survey Article: The Coming of Age of Deliberative Democracy', *The Journal of Political Philosophy*, 6(4): 400–25.

Bohman, J. (1999) 'International Regimes and Democratic Governance: Political Equality and Influence in Global Institutions', *International Affairs*, 75(3): 499–513.

Bohman, J. (2005) 'From *Demos* to *Demoi*: Democracy across Borders', *Ratio Juris*, 18(3): 293–314.

Bohman, J. (2007) 'Democratizing the Transnational Polity: The European Union and the Presuppositions of Democracy' in Eriksen. E.O. (ed.) 'How to Reconstitute Democracy in Europe? Proceedings from the RECON Opening Conference', *RECON Report* No. 3/*Arena Report* No. 8/07, pp. 65–89, www.reconproject.eu/main.php/RECONreport0307.pdf?fileitem=4473010 (25 May 2012).

Bohman, J. and Rehg, W. (eds) (1997) *Deliberative Democracy. Essays on Reason and Politics*, Cambridge, MA and London: MIT Press.

Bostock, D. (2002) 'Coreper Revisited', *Journal of Common Market Studies*, 40(2): 215–34.

Bovens, M. (2007) 'Analysing and Assessing Accountability: A Conceptual Framework', *European Law Journal*, 13(4): 447–68.

Bovens, M., Curtin, D. and t'Hart, P. (eds) (2010) *The Real World of EU Accountability: What Deficit?*, Oxford: Oxford University Press.

Bowler, S. and Farrell, D (1995) 'The Organizing of the European Parliament: Committees, Specialization and Co-ordination', *British Journal of Political Science*, 2(2): 219–43.

Boyron, S. (1996) 'Maastricht and the Codecision Procedure: A Success Story', *International & Comparative Law Quarterly*, 45(2): 293–318.

Broscheid, A. and Coen, D. (2003) 'Insider and Outsider Lobbying of the European Commission. An Informational Model of Forum Politics', *European Union Politics*, 4(2): 165–89.

Bunyan, T. (2000) 'Access to Documents "Could Fuel Public Discussion" ' in Bunyan, T., Curtin, D. and White, A., 'Essays for an Open Europe' European Federation of Journalists, www.statewatch.org/secret/essays.pdf (25 May 2012).

Bunyan, T. (2005) 'More Openness or Just a Drop in the Ocean? The Need for Freedom of Information in the EU', Statewatch, www.statewatch.org/news/2005/dec/03foi-in-eu.htm (25 May 2012).

Bunyan, T. (2007a) 'Secret Trilogues and the Democratic Deficit', Statewatch Viewpoint, www.statewatch.org/news/2007/sep/ep-co-decision-secret-trilogues.pdf (25 May 2012).

Bunyan, T. (2007b) 'European Parliament: Secret Trilogues and the Democratic Deficit', www.spectrezine.org/europe/Bunyan3.htm (25 May 2012).

Burns, C. (2004) 'Codecision and the European Commission: a Study of Declining Influence?', *Journal of European Public Policy*, 11(1): 1–18.

Burns, C. (2006) 'Co-decision and Inter-Committee Conflict in the European Parliament Post-Amsterdam', *Government and Opposition*, 41(2): 230–48.

Buzogány, A. and Stuchlík, A. (2011) 'Paved with Good Intentions. Ambiguities of Empowering Parliaments after Lisbon', German Research Institute For Public Administration Speyer, FÖV 66 Discussion Papers.

Cameron, F. (2007) *An Introduction to European Foreign Policy*, London and New York: Routledge.

Carubba, C.J., Gabel, M., Murrah, L., Clough, R., Montgomery, E. and Schambach, R. (2004) 'A second look at Legislative Behaviour in the European Parliament. Roll-call votes and the Party System', Political Science Series 94, Institute for Advanced Studies, Vienna, www.ihs.ac.at/publications/pol/pw_94.pdf (25 May 2012).

Chalmers, D., Hadjiemmanuil, C., Monti, G. and Tomkins, A. (2006) *European Union Law*, Cambridge: Cambridge University Press.
Chambers, S. (1996) *Reasonable Democracy. Jürgen Habermas and the Politics of Discourse*, Ithaca and London: Cornell University Press.
Chambers, S. (2003) 'Deliberative Democratic Theory', *Annual Review of Political Science*, 6: 307–26.
Chambers, S. (2004a) 'Behind Closed Doors: Publicity, Secrecy, and the Quality of Deliberation', *The Journal of Political Philosophy*, 12(4): 389–410.
Chambers, S. (2004b) 'Democracy, Popular Sovereignty, and Constitutional Legitimacy', *Constellations*, 11(2): 153–73.
Chambers, S. (2005) 'Measuring Publicity's Effect: Reconciling Empirical Research and Normative Theory', *Acta Politica*, 40(2): 255–66.
Checkel, J.T. (2001) 'Why Comply? Social Learning and European Identity Change', *International Organization*, 55(3): 553–88.
Christiansen, T. (2006a) 'The European Commission. The European Executive Between Continuity and Change' in Richardson, J. (ed.) *European Union Power and Policy-Making*, Oxon: Routledge, 3rd edition, pp. 99–120.
Christiansen, T. (2006b) Out of the Shadows: The General Secretariat of the Council of Ministers', *The Journal of Legislative Studies*, 8(4): 80–97.
Christiansen, T., Føllesdal, A. and Piattoni, S. (2003) 'Informal Governance in the European Union: an Introduction' in Christiansen, T. and Piattoni, S. (eds) *Informal Governance in the European Union*, Cheltenham, UK and Northampton, MA: Edward Elgar Publishing Ltd., pp. 1–21.
Cini, M. (2006) 'The European Commission: An Unelected Legislator?', *The Journal of Legislative Studies*, 8(4): 14–26.
Closa, C. and Fossum, J.E. (eds) (2004) 'Deliberative Constitutional Politics in the EU', *Arena Report* 05/2004.
Cohen, J. (1999) 'Reflections on Habermas on Democracy', *Ratio Juris*, 12(4): 385–416.
Cohen, J. and Sabel, C. (1997) 'Directly-Deliberative Polyarchy', *European Law Journal*, 3(4): 313–42.
Cohen, J. and Sabel, C. (2003) 'Sovereignty and Solidarity: EU and US' in Zeitlin, J. and Trubek, D.M. (eds) *Governing Work and Welfare in a New Economy. European and American Experiments*, Oxford: Oxford University Press, pp. 345–75.
Collins, K., Burns, C. and Warleigh, A. (1998) 'Policy Entrepreneurs: The Role of European Parliament Committees in the Making of EU Policy', *Statute Law Review*, 19(1): 1–11.
Cooper, I. (2012) 'A "Virtual Third Chamber" for the European Union? National Parliaments after the Treaty of Lisbon', *West European Politics*, 35(3): 441–65.
Corbett, R., Jacobs, F. and Shackleton, M. (2007) *The European Parliament*, 7th edition, London: John Harper Publishing.
Cortes, C. (2000) 'Co-decision: The Recent Developments', www.europeum.org/doc/obrazky/cislo5/english.htm (25 May 2012).
Costa, O. (2003) 'The European Court of Justice and Democratic Control in the European Union', *Journal of European Public Policy*, 10(5): 740–61.
Crombez, C. (1997) 'The Co-decision Procedure in the European Union', *Legislative Studies Quarterly* 22 (1): 199–228.
Crombez, C. (2000) 'The Institutional Reform and Codecision in the European Union', *Constitutional Political Economy*, 11: 41–57.
Curtin, D. (1997) 'Postnational Democracy. The European Union in Search of a Political

Philosophy', Revised and Expanded Version of the Inaugural Lecture Pronounced by Dr. Deirdre M. Curtin on Accepting the Chair of the Law of International Organisations, University of Utrecht, the Netherlands, The Hague, London and Boston: Kluwer Law International.

Curtin, D. (2000) 'Citizens' Fundamental Right of Access to EU Information: An Evolving Digital *Passepartout*?', *Common Market Law Review*, 37(1): 7–41.

Curtin, D. (2001) 'Authoritarian Temptation Seduces EU Decision-Makers', http://lausti.com/focus/2001/curtin.html (25 May 2012).

Curtin, D. (2003) 'Digital Government in the European Union: Freedom of Information Trumped by "Internal Security"' in *National Security and Open Government: Striking the Right Balance*, Commentaries edited by the Campbell Public Affairs Institute from a Symposium Co-organised with the Open Society Justice Initiative, Syracuse, New York: Campbell Public Affairs Institute, pp. 101–21.

Curtin, D. (2007a) 'Holding (Quasi-)Autonomous EU Administrative Actors to Public Account', *European Law Journal*, 13(4): 523–41.

Curtin, D. (2007b) 'Transparency, Audiences and the Evolving Role of the EU Council of Ministers' in Fossum, J.E. and Schlesinger, P. (eds) *The European Union and the Public Sphere. A Communicative Space in the Making?* London and New York: Routledge, pp. 246–58.

Curtin, D. (2009) *Executive Power in the European Union. Law, Practices and the Living Constitution*, Oxford: Oxford University Press.

Davis, R.W. (1999) 'Public Access to Community Documents: a Fundamental Human Right?', *European Integration online Papers* (EIoP), 3(8), http://eiop.or.at/eiop/texte/1999–008a.htm (25 May 2012).

De Clerck-Sachsse, J. and Kaczynski, P.M. (2009) 'The European Parliament – More Powerful, Less Legitimate? An Outlook for the 7th Term' *CEPS Working Document* No. 314/May 2009.

De Leeuw, M.E. (2003) 'The Regulation on Public Access to European Parliament, Council and Commission Documents in the European Union: Are Citizens Better Off?', *European Law Review*, 28(3): 324–48.

Dewey, J. (1927) *The Public and its Problems*, London: Allen & Unwin.

Downs, A. (1957) *An Economic Theory of Democracy*, New York: Harper & Row.

Driessen, B. (2005) 'The Council of the European Union and Access to Documents', *European Law Review*, 30(5): 675–96.

Duke, S. and Vanhoonacker, S. (2006) 'Administrative Governance in the CFSP: Development and Practice', *European Foreign Affairs Review*, 11: 163–82.

Dyrberg, P. (2002) 'Accountability and Legitimacy: What is the Contribution of Transparency?' in Arnull, A. and Wincott, D. (eds) *Accountability and Legitimacy in the European Union*, Oxford: Oxford University Press, pp. 81–96.

Egeberg, M. (2003) 'The European Commission' in Cini, M. (ed.) *European Union Politics*, Oxford: Oxford University Press, pp. 131–47.

Egeberg, M., Schaefer, G.F. and Trondal, J. (2006) 'EU Committee Governance between Intergovernmental and Union Administration' in Egeberg, M. (ed.) *Multilevel Union Administration. The Transformation of Executive Politics in Europe*, Hampshire and New York: Palgrave Macmillan, pp. 66–85.

Elster, J. (ed.) (1998a) *Deliberative Democracy*, Cambridge: Cambridge University Press.

Elster, J. (1998b) 'Introduction' in Elster, J. (ed.) *Deliberative Democracy*, Cambridge: Cambridge University Press, pp. 1–18.

Elster, J. (1998c) 'Deliberation and Constitution-Making' in Elster, J. (ed.) *Deliberative Democracy*, Cambridge: Cambridge University Press, pp. 97–122.

Eriksen, E.O. (ed.) (1995) *Deliberativ Politikk. Demokrati i Teori og Praksis*, Oslo: Tano.

Eriksen, E.O. (2001) *Demokratiets Sorte Hull – om Spenningen Mellom Fag og Politikk i Velferdsstaten*, Oslo: Abstrakt Forlag.

Eriksen, E.O. (2003) 'Integration and the Quest for Consensus – On the Micro-foundation of Supranationalism' in Eriksen, E.O., Joerges, C. and Neyer, J. (eds) 'European Governance, Deliberation and the Quest for Democratisation', *Arena Report* 02/2003, pp. 159–225.

Eriksen, E.O. (2005) 'Reflexive Integration in Europe' in Eriksen, E.O. (ed.) *Making the European Polity. Reflexive integration in the EU*, London and New York: Routledge, pp. 9–29.

Eriksen, E.O. (2007a) 'Democratic Legitimacy – Working Agreement or Rational Consensus?' in Gilje, N. and Grimen, H. (eds) *Discursive Modernity. Festschrift to Professor Gunnar Skirbekk on the Occasion of his Seventieth Birthday*, Oslo: Universitetsforlaget, pp. 92–115.

Eriksen, E.O. (2007b) 'Deliberation und Demokratische Legitimität in der EU – Zwischen Konsens und Kompromiss' in Niesen, P. and Herborth, B. (eds), *Anarchie der Kommunkativen Freihet. Jürgen Habermas und die Theorie der Internationalen Politik*, Frankfurt am Main: Suhrkamp Taschenbuch Verlag, pp. 294–320.

Eriksen, E.O. (2009) *The Unfinished Democratization of Europe*, Oxford: Oxford University Press.

Eriksen, E.O. and Fossum, J.E. (eds) (2000) *Democracy in the European Union. Integration through Deliberation*, London and New York: Routledge.

Erikson, E.O. and Fossum, J.E. (eds) (2002), 'Democracy Through Strong Publics in the European Union?', *Journal of Common Market Studies*, 40(3): 401–24.

Eriksen, E.O. and Neyer, J. (2003) 'The Forging of Deliberative Supranationalism in the EU?' in Eriksen, E.O., Joerges, C. and Neyer, J. (eds) 'European Governance, Deliberation and the Quest for Democratisation', *Arena Report* 02/2003, pp. 1–22.

Eriksen, E.O. and Skivenes, M.I. (2000) 'Nye Deltagelsesformer og Demokratisk Medvirkning', LOS-senter Rapport 0010: Bergen.

Eriksen, E.O. and Weigård, J. (1999) *Kommunikativ Handling og Deliberativt Democrati. Jürgen Habermas' Teori om Politikk og Samfunn*, Bergen: Fagbokforlaget.

Eriksen, E.O. and Weigård, J. (2003) *Understanding Habermas. Communicative Action and Deliberative Democracy*, London: Continuum Press.

Estlund, D. (1997) 'Beyond Fairness and Deliberation: The Epistemic Dimension of Democratic Authority' in Bohman, J. and Rehg, W. (eds) *Deliberative Democracy. Essays on Reason and Politics*, Cambridge, MA and London, England: MIT Press, pp. 173–204.

Farrell, Henry and Héritier, Adrienne (2003a) 'Formal and Informal Institutions Under Codecision: Continuous Constitution-Building in Europe', *Governance*, 16(4): 577–600.

Farrell, Henry and Héritier, Adrienne (2003b) 'The Invisible Transformation of Codecision: Problems of Democratic Legitimacy', *Sieps Report* No. 7, June 2003, Swedish Institute for European Policy Studies.

Farrell, Henry and Héritier, Adrienne (2004) 'Interorganizational Negotiations and Intraorganizational Power in Shared Decision Making. Early Agreements under Codecision and their Impact on the European Parliament and Council', *Comparative Political Studies*, 37(10): 1184–1212.

Farrell, Henry and Héritier, Adrienne (2007) 'Codecision and Institutional Change', *West European Politics*, 30(2): 285–300.

Ferrara, F. and Weishaupt, J.T. (2004) 'Get Your Act Together. Party Performance in European Parliament Elections', *European Union Politics*, 5(3): 283–306.

Fishkin, J. (1999) 'Toward a Deliberative Democracy: Experimenting with an Ideal' in Elkin, S. and Soltan, K.E. (eds) *Citizen Competence and Democratic Institutions*, State College, PA: State University Press, pp. 279–90.

Fishkin, J. and Luskin, R.C. (2005) 'Experimenting with a Democratic Ideal: Deliberative Polling and Public Opinion', *Acta Politica*, 40(3): 284–98.

Fouilleux, E., de Maillard, J. and Smith, A. (2005) 'Technical or Political? The Working Groups of the EU Council of Ministers', *Journal of European Public Policy*, 12(4): 609–23.

Forst, R. (2001) 'The Rule of Reasons. Three Models of Deliberative Democracy', *Ratio Juris*, 14(4): 345–78.

Fossum, J.E. and Menéndez, A.J. (2005) 'The Constitution's Gift? A Deliberative Democratic Analysis of Constitution Making in the European Union', *European Law Journal*, 11(4): 380–410.

Fraser, N. (1992) 'Rethinking the Public Sphere. A Contribution to the Critique of Actually Existing Democracy' in Calhoun, C. (ed.) *Habermas and the Public Sphere*, Cambridge, MA: MIT Press, pp. 109–42.

Fraser, N. (2005) 'Reframing Justice in a Globalizing World', *New Left Review*, 36 November 2005, pp. 69–88.

Føllesdal, A. and Hix, S. (2006) 'Why there is a Democratic Deficit in the EU: A Response to Majone and Moravcsik', *Journal of Common Market Studies*, 44(3): 533–62.

Fung, A. and Wright, E.O. (2001) 'Deepening Democracy: Innovations in Empowered Participatory Governance', *Politics & Society*, 29(1): 5–41.

Garman, J. and Hilditch, L. (1998) 'Behind the Scenes: an Examination of the Importance of the Informal Processes at Work in Conciliation', *Journal of European Public Policy*, 5(2): 271–84.

Garrett, G., Tsebelis, G. and Corbett, R. (2001) 'Forum section: The EU Legislative Process. Academics vs. Practitioners – Round 2', *European Union Politics*, 2(3): 353–66.

Gaus, D. (2008) 'Legitimate Political Rule Without a State? An Analysis of Joseph H.H. Weiler's Justification of the Legitimacy of the European Union Qua Non-Statehood' *RECON Online Working Papers* 2008/12.

Gehring, T. and Kerler, M. (2008) 'Institutional Stimulation of Deliberative Decision-Making: Division of Labour, Deliberative Legitimacy and Technical Regulation in the European Single Market', *Journal of Common Market Studies*, 46(5): 1001–23.

Gerstenberg, O. (1997) 'Law's Polyarchy: a Comment on Sabel and Cohen', *European Law Journal*, 3(4): 343–58.

Gerstenberg, O. and Sabel, C. (2002) 'Directly-Deliberative Polyarchy: An Institutional Ideal for Europe?' in Joerges, C. and Dehousse, R. (eds) *Good Governance in Europe's Integrated Market*, Oxford: Oxford University Press, pp. 289–341.

Goodin, R. (2005) 'Sequencing Deliberative Moments', *Acta Politica*, 40(2): 182–96.

Goodin, R. and Niemeyer, S.J. (2003) 'When Does Deliberation Begin? Internal Reflection versus Public Discussion in Deliberative Democracy', *Political Studies*, 51: 627–49.

Gornitzka, Å. and Sverdrup, U. (2008) 'Who Consults? The Configuration of Expert Groups in the European Commission', *West European Politics*, 3(4): 725–50.

Greenwood, J. (2003) *Interest Representation in the European Union*. The European Union Series, Houndsmill, Basingstoke, Hampshire and New York: Palgrave Macmillan.
Gutmann, A. and Thompson, D. (2002) 'Deliberative Democracy Beyond Process', *The Journal of Political Philosophy*, 10(2): 153–74.
Gutmann, A. and Thompson, D. (2004) *Why Deliberative Democracy?* Princeton and Oxford: Princeton University Press.
Habermas, J. (1981a) *The Theory of Communicative Action: Volume One. Reason and the Rationalization of Society*, Boston, MA: Beacon Press.
Habermas, J. (1981b) *The Theory of Communicative Action: Volume Two. Lifeworld and System: A Critique of Functionalist Reason*, Boston, MA: Beacon Press.
Habermas, J. (1998) *Between Facts and Norms. Contributions to a Discourse Theory of Law and Democracy*, Cambridge, MA: MIT Press.
Habermas, J. (2001) *The Postnational Constellation. Political Essays*, Cambridge, MA: MIT Press.
Habermas, J. (2006) 'Political Communication in Media Society: Does Democracy Still Enjoy an Epistemic Dimension? The Impact of Normative Theory on Empirical Research', *Communication Theory*, 16: 411–26.
Häge, F.M. and Kaeding, M. (2007) 'Reconsidering the European Parliament's Legislative Influence: Formal vs. Informal Procedures', *European Integration*, 29(3): 341–61.
Hagemann, S. (2009) 'Strength in Numbers? An Evaluation of the 2004–2009 European Parliament', *EPC Issue Paper* No. 58, May 2009, European Policy Centre, pp. 1–26.
Hagemann, S. and De Clerck-Sachsse, J. (2007) 'Old Rules, New Game. Decision-Making in the Council of Ministers after the 2004 Enlargement', Special Report March 2007, Centre for European Policy Studies (CEPS).
Hardacre, A. and Kaeading, M. (2011) 'Delegated and Implementing Acts. The New Comitology', EIPA Essential Guide, Version 2, March 2011, www.eipa.eu/files/repository/product/20110902144617_Comitology_Brochure3EN_web.pdf (25 May 2012).
Hausemer, P. (2006) 'Participation and Political Competition in Committee Report Allocation. Under What Conditions Do MEPs Represent Their Constituents?', *European Union Politics*, 7(4): 505–30.
Hayes, B. (2005) 'The Right to Know or the Right to Find Out? The Need for an EU Freedom of Information Law' Statewatch, www.statewatch.org/news/2005/nov/eu-FOI.pdf (25 May 2012).
Hayes-Renshaw, F. and Wallace, H. (2003) 'Reforming the Council: A Work in Progress' Sieps Report No. 9, June 2003, Swedish Institute for European Policy Studies.
Hayes-Renshaw, F. and Wallace, H. (2006) *The Council of Ministers*, second edition, Houndmills, Basingstoke, Hampshire and London: Macmillan Press Ltd.
Hayes-Renshaw, F., van Aken, W. and Wallace, H. (2005) 'When and Why the Council of Ministers of the EU Votes Explicitly', *EUI Working Papers*, RSCAS No. 2005/25, www.iue.it/RSCAS/WP-Texts/05_25.pdf (25 May 2012).
Heisenberg, D. (2005) 'The Institution of "Consensus" in the European Union: Formal versus Informal Decision-Making in the Council', *European Journal of Political Research*, 44: 65–90.
Heliskoski, J. and Leino, P. (2007) 'Darkness at Break of Noon: The Case Law on Regulation No. 1049/2001 on Access to Documents', *Common Market Law Review*, 43: 735–81.
Héritier, A. (2003) 'Composite Democracy in Europe: the Role of Transparency and Access to Information', *Journal of European Public Policy*, 10(5): 814–33.

Héritier, A. (2012) 'Institutional Change in Europe: Co-decision and Comitology Transformed', *Journal of Common Market Studies*, 50(S1): 38–54.

Héritier, A. and Moury, C. (2011) 'Contested Delegation: The Impact of Co-decision on Comitology', *West European Politics*, 34(1): 145–66.

Héritier, A. and Reh, C. (2011) 'Codecision and its Discontents: Intra-organizational Politics and Institutional Reform in the European Parliament' Paper Presented for the EUSA Biannual Meeting, Boston, 3–5 March 2011, www.ucl.ac.uk/silva/spp/research/esrc-project/Paper_EUSA_2011_Heritier-Reh.pdf (25 May 2012).

Hix, S. (2002) 'Constitutional Agenda-Setting Through Discretion Rule Interpretation: Why the European Parliament Won at Amsterdam', *British Journal of Political Science*, 32(2): 259–80.

Hix, S. (2005) *The Political System of the European Union*, 2nd edition, Basingstoke: Palgrave, Macmillan.

Hix, S., Noury, A. and Roland, G. (2005) 'Power to the Parties: Cohesion and Competition in the European Parliament, 1979–2001', *British Journal of Political Science* 35(2): 209–34.

Hix, S., Noury, A. and Roland, G. (2006) 'Dimensions of Politics in the European Parliament', *American Journal of Political Science*, 50(2): 494–511.

Holzinger, K. (2005) 'Context or Conflict Types: Which Determines the Selection of Communication Mode', *Acta Politica*, 40(2): 239–54.

Høyland, B. (2006) 'Allocations of Codecision. Reports in the Fifth European Parliament', *European Union Politics*, 7(1): 30–50.

Høyland, B. and Hagemann, S. (2010) 'Bicameral Politics in the European Union', *Journal of Common Market Studies*, 48(4): 811–33.

Hubschmid, C. and Moser, P. (1997) 'The Cooperation Procedure in the EU: Why Was the European Parliament Influential in the Decision about Car Emissions Standards?', *Journal of Common Market Studies*, 35(2): 225–41.

Joerges, C. (2006) 'Deliberative Political Processes Revisited: What Have We Learnt About the Legitimacy of Supranational Decision-Making?', *Journal of Common Market Studies*, 44(4): 779–802.

Joerges, C. and Neyer, J. (1997) 'From Intergovernmental Bargaining to Deliberative Political Processes. The Constitutionalization of Comitology', *European Law Journal*, 3(3): 273–99.

Joerges, C. and Vos, E. (1999) *EU Committees: Social Regulation, Law and Politics*, Oxford, Portland: Hart Publishing.

Judge, D. and Earnshaw, D. (2003) *The European Parliament*, Houndmills, Basingstoke, Hampshire: Palgrave, Macmillan.

Kaeding, M. (2005) 'The World of Committee Reports: Rapporteurship Assignment in the European Parliament', *The Journal of Legislative Studies*, 11(1): 82–104.

Kasack, C. (2004) 'The Legislative Impact of the European Parliament under the Revised Co-Decision Procedure. Environmental, Public Health and Consumer Policies', *European Union Politics*, 5(2): 241–60.

Keohane, R.O., Macedo, S. and Moravcsik, A. (2009) 'Democracy-Enhancing Multilateralism', *International Organization*, 63(1): 1–31.

Kiiver, P. (2011) 'The Early-Warning System for the Principle of Subsidiarity: the National Parliaments as a Conseil d'Etat for Europe', *European Law Review*, 36(1): 98–108.

Kohler-Koch, B. and Finke, B. (2007) 'The Institutional Shaping of EU-Society Relations: A Contribution to Democracy via Participation?', *Journal of Civil Society*, 3(3): 205–21.

Kohler-Koch, B. and Maloney, W. (eds) (2008) 'Opening EU-Governance in Civil Society. Gains and Challenges', *CONNEX Report Series* No. 5, www.mzes.uni-mannheim.de/projekte/typo3/site/fileadmin/BookSeries/Volume_Five/CONNEX%20Report%20Series_Book%205.pdf (25 May 2012).

Kohler-Koch, B. and Rittberger, B. (eds) (2007) *Debating the Democratic Legitimacy of the European Union*, US: Rowman and Littlefield Publishers.

König, T., Lindberg, B., Lechner, S. and Pohlmeier, W. (2007) 'Bicameral Conflict Resolution in the European Union: An Empirical Analysis of Conciliation Committee Bargains', *British Journal of Political Science*, 37 (2): 281–312.

Kranenborg, H. and Voermans, W. (2005) *Access to Information in the European Union. A Comparative Analysis of EC and Member State Legislation*, Groningen/Amsterdam: Europa Law Publishing.

Kreppel, A. (2002) 'Moving Beyond Procedure. An Empirical Analysis of European Parliament Legislative Influence', *Comparative Political Studies*, 35(7): 784–813.

Laming, R. (2006) 'Openness and Secrecy in the EU Institutions: Lessons from the EU Sugar Regime', Policy Brief No. 28, 5 June 2006, The Federal Trust, www.fedtrust.co.uk/admin/uploads/PolicyBrief28.pdf (25 May 2012).

Landwehr, C. (2010) 'Discourse and Coordination: Modes of Interaction and their Roles in Political Decision-Making', *The Journal of Political Philosophy*, 18(1): 101–22.

Larsson, T. and Murk, J. (2007) 'The Commission's Relations with Expert Advisory Groups' in Christiansen, T. and Larsson, T (eds) *The Role of Committees in the Policy-Process of the European Union. Legislation, Implementation and Deliberation*, Cheltenham, UK, Northhampton, MA: Edward Elgar, pp. 64–95.

Lewis, J. (1998) 'Is the "Hard Bargaining" Image of the Council Misleading? The Committee of Permanent Representatives and the Local Elections Directives', *Journal of Common Market Studies*, 36(4): 479–504.

Lewis, J. (2000) 'The Methods of Community in EU Decision-Making and Administrative Rivalry in the Council's Infrastructure', *Journal of European Public Policy*, 7(2): 261–89.

Lord, C. (2004) *A Democratic Audit of the European Union*, Houndsmill, Basingstoke, Hampshire and New York: Palgrave Macmillan.

Lord, C. (2007a) 'Parliamentary Representation in a Decentered Polity' in Kohler-Koch, B. and Rittberger, B. (eds) *Debating the Democratic Legitimacy of the European Union*, Lanham: Rowman and Littlefield, pp. 139–56.

Lord, C. (2007b) 'Contested Meanings, Democracy Assessment and the European Union', *Comparative European Politics*, 5: 70–86.

MacIntyre, A. (1981) *After Virtue*, London: Duckworth.

MacIntyre, A. (1998) *Whose Justice, Which Rationality?* London: Duckworth.

Magnette, P. (2003) 'Between Parliamentary Control and the Rule of Law: the Political Role of the Ombudsman in the European Union', *Journal of European Public Policy*, 10(5): 677–94.

Magnette, P. (2004) 'When Does Deliberation Matter? Constitutional Rhetoric in the Convention on the Future of Europe' in Closa, C. and Fossum, J.E. (eds) 'Deliberative Constitutional Politics in the EU', *Arena Report* 05/2004, pp. 205–40.

Mahoney, C. (2004) 'The Power of Institutions. State and Interest Group Activity in the European Union', *European Union Politics*, 5(4): 441–66.

Majone, G. (1996) *Regulating Europe*, London: Routledge.

Mamadouh, V. and Raunio, T. (2003) 'The Committee System: Powers, Appointments and Report Allocation', *Journal of Common Market Studies*, 41(2): 333–51.

Mansbridge, J. (1992) 'A Deliberative Theory of Interest Representation' in Petracca, M.P. (ed.) *The Politics of Interests. Interest Groups Transformed*, Boulder, San Francisco, Oxford: Westview Press, pp. 32–57.

Mansbridge, J. (2003) 'Rethinking Representation', *American Political Science Review*, 97(4): 515–28.

Mansbridge, J. with Bohman, J., Chambers, S., Estlund, D., Føllesdal, A., Fung, A., Lafont, C., Manin, B. and Martí, J.L. (2010) 'The Place of Self-Interest and the Role of Power in Deliberative Democracy', *Journal of Political Philosophy*, 18(1): 64–100.

March, J. and Olsen, J.P. (1998) 'The Institutional Dynamics of International Political Order', *International Organization*, 52(4): 943–69.

March, J. and Olsen, J.P. (1989) *Rediscovering Institutions*, New York: Free Press.

March, J. and Olsen, J.P. (2006a) 'The Logic of Appropriateness' in Rein, M., Moran, M. and Goodin, R.R. (eds) *The Oxford Handbook of Public Policy*, Oxford: Oxford University Press, pp. 689–708.

March, J. and Olsen, J.P. (2006b) 'Elaborating the "New Institutionalism"', in Rhodes, R.A.W. Binder, S. and Rockman, B. (eds) *The Oxford Handbook of Political Institutions*, Oxford: Oxford University Press, pp. 3–20.

Mattila, M. (2004) 'Contested Decisions: Empirical Analysis of Voting in the European Union Council of Ministers', *European Journal of Political Research*, 43: 29–50.

Maurer, A. (2003) 'The Legislative Powers and Impact of the European Parliament', *Journal of Common Market Studies*, 41(2): 227–47.

Mazey, S. and Richardson, J. (2006a) 'Interest Groups and EU Policy-Making. Organisational Logic and Venue Shopping' in Richardson, J. (ed.) *European Union Power and Policy-Making*, 3rd edition, Oxon: Routledge, pp. 247–68.

Mazey, S. and Richardson, J. (2006b) 'The Commission and the Lobby' in Spence, D. (ed.) with Edwards, G., *The European Commission*, London: John Harper Publishing, pp. 279–92.

McElroy, G. (2006) 'Committee Representation in the European Parliament', *European Union Politics*, 7(1): 5–29.

Menéndez, A.J. (2003) 'Three Conceptions of the Constitution' *ARENA Working Paper Series*, 12/2003.

Menéndez, A.J. (2009) 'The European Democratic Challenge: The Forging of a Supranational *Volonté Générale*', *European Law Journal*, 15(3): 277–308.

Mill, J.S. (1861) *Considerations on Representative Government*, [reprint, 1984], London: Dent.

Miller, D. (1995) *On Nationality*, Oxford: Clarendon Press.

Miller, D. (2000) *Citizenship and National Identity*, Cambridge: Polity Press.

Miller, D. (2009) 'Democracy's Domain', *Philosophy & Public Affairs*, 37(3): 201–28.

Moravcsik, A. (2002) 'In Defence of the "Democratic Deficit": Reassessing Legitimacy in the European Union', *Journal of Common Market Studies*, 40(4): 603–24.

Moser, P. (1996), 'The European Parliament as a Conditional Agenda-Setter: What are the Conditions? A Critique of Tsebelis', *American Political Science Review*, 90(4): 834–8.

Nanz, P. and Steffek, J. (2005) 'Assessing the Democratic Quality of Deliberation in International Governance: Criteria and Research Strategies', *Acta Politica*, 40(3): 368–83.

Napel, S. and Widgrén, M. (2006) 'The Inter-institutional Distribution of Power in the EU Codecision', *Social Choice and Welfare*, 27: 129–54.

Naurin, Daniel (2007) *Deliberation Behind Closed Doors. Transparency and Lobbying in the European Union*, Colchester: ECPR Press.

Naurin, D. and Rasmussen, A. (2011) 'New External Rules, New Internal Games: How the EU Institutions Respond when Inter-institutional Rules Change', *West European Politics*, 34(1): 1–17.

Neblo, M.A. (2005) 'Thinking Through Democracy: Between the Theory and Practice of Deliberative Politics', *Acta Politica*, 40(2): 169–81.

Neuhold, C. (2001) 'Into the New Millennium: The Evolution of the European Parliament from Consultative Assembly to Co-legislator', *Eipascope* 2001/1, www.eipa.eu/files/repository/eipascope/2000_1_1.pdf (25 May 2012).

Neuhold, C. (2007) 'Achieving Consensus Within European Parliament Committees: How Does the Parliament (Continue to) Manage' Draft Paper for CONNEX Conference, Research Group 1, Barcelona 7–9th June 2007.

Neuhold, C. and Settembri, P. (2008) 'Achieving Consensus Through Committees: Does the European Parliament Manage?', *Journal of Common Market Studies*, 47(1): 127–51.

Neunreither, K. (2002) 'Elected Legislators and their Unelected Assistants in the European Parliament', *The Journal of Legislative Studies*, 8(4): 40–60.

Nugent, N. (2001) *The European Commission*, Houndmills, Basingstoke, Hampshire and New York: Palgrave.

Öberg, U. (1998) 'Public Access to Documents After the Entry into Force of the Amsterdam Treaty: Much Ado About Nothing?', *European Integration Online Papers* (EIoP), 2(8), http://eiop.or.at/eiop/pdf/1998–008.pdf (25 May 2012).

Öberg, U. (1999) 'EU Citizens' Right to Know: the Improbable Adoption of a European Freedom of Information Act', *Cambridge Yearbook of European Legal Studies*, 2: 303–28.

O'Brennan, J. and Raunio, T. (2007) 'Introduction' in O'Brennan and Raunio (eds) *National Parliaments within the Enlarged European Union*, Oxon: Routledge, pp. 1–26.

Olsen, J.P. (2007) *Europe in Search of Political Order. An institutional perspective on unity/diversity, citizens/their helpers, democratic design/historical drift and the co-existence of orders*, Oxford, New York: Oxford University Press.

Peers, S. (2002) 'The New Regulation on Access to Documents: A Critical Analysis', *Queen's Papers on Europeanisation*, No. 6/2002, www.qub.ac.uk/schools/SchoolofPoliticsInternationalStudiesandPhilosophy/FileStore/EuropeanisationFiles/Filetoupload,38419,en.pdf (25 May 2012).

Peers, S. (2006) 'Openness and transparency in the Council', Statewatch Analysis 20 June 2006, www.statewatch.org/news/2006/jun/statewatch-analysis-openness.pdf (25 May 2012).

Pollack, M.A. (2000) 'The End of Creeping Competence? EU Policy-Making Since Maastricht', *Journal of Common Market Studies*, 38(3) 519–38.

Quittkat, C. and Finke, B. (2008), 'The EU Commission Consultation Regime' in Kohler-Koch, B. and Maloney, W. (eds) 'Opening EU-Governance in Civil Society. Gains and Challenges' *CONNEX Report Series* No. 5, pp. 183–222, www.mzes.uni-mannheim.de/projekte/typo3/site/fileadmin/BookSeries/Volume_Five/CONNEX%20Report%20Series_Book%205.pdf (25 May 2012).

Rasmussen, A. (2003) 'The Role of the European Commission in Co-decision – A Strategic Facilitator Operating in a Situation of Structural Disadvantage', *European Integration online Papers* (EIoP), 7(10), http://eiop.or.at/eiop/pdf/2003–010.pdf (25 May 2012).

Rasmussen, A. (2005) 'EU Conciliation Delegates: Responsible or Runaway Agents?', *West European Politics*, 28(5): 1015–34.

Rasmussen, A. (2007) 'Early Conclusion in the Co-decision Legislative Procedure', *EUI Working Papers*, MWP 2007/31, pp. 1–16.

Rasmussen, A. (2008) 'The EU Conciliation Committee. One or Several Principals?', *European Union Politics*, 9(1): 87–113.

Rasmussen, A. and Shackleton, M. (2005) 'The Scope for Action of European Parliament Negotiators in the Legislative Process: Lessons of the Past and for the Future', Paper prepared for the Ninth Biennial International Conference of the European Union Studies Association, Austin, Texas, March 31-April 2, 2005, http://aei.pitt.edu/2983/02/EUSA_Rasmussen_and_Shackleton1.doc (25 May 2012).

Rawls, J. (1971) *A Theory of Justice*. Cambridge, Massachusets: Harvard University Press.

Rawls, J. (1993) *Political Liberalism*. New York: Columbia University Press.

Reh, C. (2008) 'The Informal Politics of Co-Decision: Towards A Normative Assessment', Paper prepared for the UACES Conference on *Exchanging Ideas on Europe*, University of Edinburgh, 1–3 September 2008, www.ucl.ac.uk/silva/spp/research/esrc-project/Paper_UACES_Reh.pdf (25 May 2012).

Reh, C., Héritier, A., Bressanelli, E. and Koop, C. (forthcoming 2013) 'The Informal Politics of Legislation: Explaining Secluded Decision-Making in the European Union', *Comparative Political Studies*, Published before print 5 December 2011.

Reif, K. and Schmitt, H. (1980) 'Nine Second-Order Elections: a Conceptual Framework Analysis of European Election Results', *European Journal of Political Research*, 8: 3–44.

Renan, E. (1882) 'What is a Nation?' in Zimmern, A. (ed.) *Modern Political Doctrines* [reprint, 1939], London and New York: Oxford University Press.

Richardson, J. (ed.) (2006) *European Union. Power and policy-making*, Oxon and New York: Routledge.

Rhinard, M. (2002) 'The Democratic Legitimacy of the European Union Committee System', *Governance*, 15(2): 185–210.

Riker, W.H. (1982) *Liberalism Against Populism: A Confrontation Between the Theory of Democracy and the Theory of Social Choice*, San Francisco: Freeman.

Rittberger, B. (2000) 'Impatient Legislators and New Issue-Dimensions: a Critique of the Garrett-Tsebelis "Standard Version" of legislative politics', *Journal of European Public Policy*, 7(4): 554–75.

Rittberger, B. (2005) *Building Europe's Parliament. Democratic Representation Beyond the Nation-State*, Oxford: Oxford University Press.

Rittberger, B. (2007) 'Constructing Parliamentary Democracy in the European Union: How Did It Happen?' in Kohler-Koch, B. and Rittberger, B. (eds) *Debating the Democratic Legitimacy of the European Union*, US: Rowman and Littlefield Publishers, pp. 111–37.

Rohrschneider, R. and Clark, N. (2008) 'Second-Order Elections versus First-Order Thinking: How Voters Perceive the Representation Process in a Multi-Layered System of Governance' in van der Eijk, C. and Schmitt, H. (eds) 'The Multilevel Electoral System of the EU', *CONNEX Report Series*, No. 4, pp. 137–62.

Rosenberg, Shawn W. (ed.) (2007) *Deliberation, Participation and Democracy. Can the People Govern?* Houndmills, Basingstoke, Hampshire and New York: Palgrave Macmillan.

Schmalz-Bruns, R. (1999) 'Deliberativer Suprantionalismus', *Zeitschrift für Internationale Beziehungen*, 6(2): 185–242.

Schmitt, H. (2005) 'The European Parliament Elections of June 2004: Still Second-Order?', *West European Politics*, 28(3): 650–79.

Schneider, G. and Baltz, K. (2005) 'Domesticated Eurocrats: Bureaucratic Discretion in the Legislative Pre-Negotiations of the European Union', *Acta Politica*, 40 (1): 1–27.
Schumpeter, J.A. (1942) *Capitalism, Socialism and Democracy*, New York: Harper and Brothers.
Scully, R. (1997a) 'The European Parliament and the Co-decision Procedure: A Reassessment', *The Journal of Legislative Studies*, 3(3): 58–97.
Scully, R. (1997b) 'The European Parliament and the Co-decision Procedure: A Rejoinder to Tsebelis and Garrett', *The Journal of Legislative Studies*, 3(3): 93–103.
Scully, R. (1997c) 'Positively My Last Word on Co-Decision', *The Journal of Legislative Studies*, 3(4) 144–6.
Selck, T. and Steunenberg, B. (2004) 'Between Power and Luck. The European Parliament in the EU Legislative Process', *European Union Politics*, 5(1): 25–46.
Settembri, P. (2004) 'When is a Group not a Political Group? The Dissolution of the TDI Group in the European Parliament', *The Journal of Legislative Studies*, 10(1): 150–74.
Settembri, P. (2005) 'Transparency and the EU Legislator: "Let He Who is Without Sin Cast the First Stone"' *Journal of Common Market Studies*, 43(3): 637–54.
Settembri, P. (2006) 'Is the European Parliament competitive or consensual… "and why bother"?' Paper presented at the Conference "The European Parliament and the European Political Space" London 30 March 006, Federal Trust and Stiftung Wissenschaft und Politik, Berlin, www.fedtrust.co.uk/admin/uploads/FedT_workshop_Settembri.pdf (25 May 2012).
Settembri, P. (2007) 'The Surgery Succeeded. Has the Patient Died? The Impact of Enlargement on the European Union', *Jean Monnet Working Paper* 04/07.
Shackleton, M. (2000) 'The Politics of Codecision', *Journal of Common Studies*, 38(2): 325–42.
Shackleton, M. and Raunio, T. (2003) 'Codecision Since Amsterdam: a Laboratory for Institutional Innovation and Change', *Journal of European Public Policy*, 10(2): 171–87.
Sjursen, H. (2011) 'Not so Intergovernmental After All? On Democracy and Integration in European Foreign and Security Policy', *Journal of European Public Policy*, 18(8): 1078–95.
Smith, A. (2007) 'European Commissioners and the Prospects of a European Public Sphere. Information, Representation and Legitimacy' in Fossum, J.E. and Schlesinger, P. (eds) *The European Union and the Public Sphere. A communicative space in the making?* London and New York: Routledge, pp. 227–45.
Smith, K. E. (2003) *European Union Foreign Policy in a Changing World*, Cambridge: Polity Press
Spence, D. (2006a) 'Chapter 1: The President, the College and the Cabinets' in Spence, D. (ed.) with Edwards, G., *The European Commission*, London: John Harper Publishing, pp. 25–74.
Spence, D. (2006b) 'Chapter 4: The Directors General and the Services: Structures, Functions and Procedures' in Spence, D. (ed.) with Edwards, G., *The European Commission*, London: John Harper Publishing, pp. 128–55.
Steiner, J., Bächtiger, A., Spörndli, M. and Steenbergen, M.R. (2004) *Deliberative Politics in Action. Analysing Parliamentary Discourse*, Cambridge: Cambridge University Press.
Steunenberg, B. (1994) 'Decision-Making Under Different Institutional Arrangements: Legislation by the European Community', *Journal of Institutional and Theoretical Economics*, 50(4): 642–69.

Stie, A.E. (2002) 'Preconditions for Democracy: National Identity versus Constitutional Patriotism – A Theoretical Analysis of the German Federal Constitutional Court's Maastricht Judgement', *Arena Report* 02/2009.

Stie, A.E. (2010a) 'Co-decision – the Panacea for EU democracy?' *Arena Report* 01/2010.

Stie, A.E. (2010b) 'Decision-Making Void of Democratic Qualities? An Evaluation of the EU's Second Pillar Decision-Making Procedure', *European Integration Online Papers*, Special Issue 1, Vol. 14.

Thompson, D.F. (2008) 'Deliberative Democratic Theory and Empirical Political Science', *Annual Review of Political Science*, 11: 497–520.

Toshkov, D. and Rasmussen, A. (2012) 'Time to Decide: The Effect of Early Agreements on Legislative Duration in the EU', *European Integration online Papers*, Vol. 16, Article 11.

Tsebelis, G. (2002) *Vet Players: How Political Institutions Work*, Princeton NJ: Princeton University Press.

Tsebelis, G. and Garrett, G. (1996) 'Agenda Setting Power, Power Indices and Decision Making in the European Union', *International Review of Law and Economics*, 16(3): 345–61.

Tsebelis, G. and Garrett, G. (1997a) 'Agenda-Setting, Vetoes and the European Union's Co-decision Procedure', *The Journal of Legislative Studies*, 3(3): 74–92.

Tsebelis, G. and Garrett, G. (1997b) 'More on the Co-Decision Endgame', *Journal of Legislative Studies*, 3(4) 139–43.

Tsebelis, G. and Garrett, G. (2000) 'Legislative Politics in the European Union', *European Union Politics*, 1(1): 9–36.

Tsebelis, G. and Garrett, G. (2001) 'The Institutional Foundations of Intergovernmentalism and Supranationalism in the European Union', *International Organization*, 55(2): 357–90.

Tsebelis, G. and Kalandrakis, A. (1999) 'The European Parliament and Environmental Legislation: The Case of Chemicals', *European Journal of Political Research*, 36(1): 119–54.

Tsebelis, G., Jensen, C.B., Kalandrakis, A. and Kreppel, A. (2001) 'Legislative Procedures in the European Union: An Empirical Analysis', *British Journal of Political Science*, 31(4): 573–99.

Ulbert, C. and Risse, T. (2005) 'Deliberately Changing the Discourse: What Does Make Arguing Effective?', *Acta Politica*, 40(3): 251–67.

Van Schendelen, R. (2006) 'The In-sourced Experts', *The Journal of Legislative Studies*, 8(4): 27–39.

Van Schendelen, R. and Scully, R. (2006) 'Introduction', *The Journal of Legislative Studies*, 8(4): 1–13.

Vos, E. (2009) '50 Years of European Integration, 45 Years of Comitology' *Maastricht Faculty of Law Working Paper* 2009/3.

Walzer, M. (1983) *Spheres of Justice*, New York: Basic Books.

Walzer, M. (1994) *Thick and Thin: Moral Argument at Home and Abroad*, Notre Dame, Indiana: University of Notre Dame Press.

Walzer, M. (1996) *What it Means to be an American: Essays on the American Experience*, New York: Marsilo Publishers Corp.

Warleigh, A. (2003) 'Informal Governance: Improving EU Democracy?' in Christiansen, T. and Piattoni, S. (eds) *Informal Governance in the European Union*, Cheltenham, UK and Northampton, MA, USA: Edward Elgar Publishing Ltd.

Warren, M.E. (2007) 'Institutionalizing Deliberative Democracy' in Rosenberg, S.W. (ed.) *Deliberation, Participation and Democracy. Can the People Govern?* Houndmills, Basingstoke, Hampshire and London: Palgrave Macmillan, pp. 272–88.

Warren, M.E. (2008) 'Citizen Representatives' in Warren, M.E. and Pearse, H., *Designing Deliberative Democracy. The British Columbia Citizens' Assembly*, Cambridge: Cambridge University Press, pp. 50–69.

Warren, M.E. and Pearse, H. (eds) (2008) *Designing Deliberative Democracy. The British Columbia Citizens' Assembly*, Cambridge: Cambridge University Press.

Weiler, J.H.H. (1995a) 'The State "über alles". Demos, Telos and the German Maastricht Decision', *Jean Monnet Working Papers* 6/95.

Weiler, J.H.H. (1995b) 'Does Europe Need a Constitution? Reflections on Demos Telos and the German Maastricht Decision', *European Law Journal*, 1(3): 219–58.

Westlake, M. (2006) 'The European Commission and the European Parliament' in Spence, D. (ed.) with Edwards, G., *The European Commission*, London, UK: John Harper Publishing, pp. 263–78.

Westlake, M. and Galloway, D. (2004) 'Chapter 12. Other Council Preparatory Bodies' in Westlake, M. and Galloway, D. (eds) with Digneffe, T., 3rd edition, *The Council of the European Union*, London: John Harper Publishing, pp. 217–22.

Whitaker, R. (2001) 'Party Control in a Committee-Based Legislature? The Case of the European Parliament', *The Journal of Legislative Studies*, 7(4): 63–88.

Williamson, D. (2006) 'Conciliation Procedure in the European Law-making Process', *The Journal of Legislative Studies*, 12(1): 1–7.

Yordanova, N. (2011) 'Inter-institutional Rules and Division of Power in the European Parliament: Allocation of Consultation and Co-decision Reports', *West European Politics*, 34(1): 97–121.

Young, I.M. (1990) *Justice and the Politics of Difference*, Princeton, New Jersey: Princeton University Press.

Official EU documents

Council documents

Council of the European Union (1993) 'Council Decision 93/731/EC of 20 December 1993 on public access to Council documents' (93/731/EC), 20 December1993, http://ue.eu.int/uedocs/cmsUpload/register5.pdf (25 May 2012).
Council of the European Union (1996) 'Council Decision of 6 December 1996 amending Decision 93/731/EC on public access to Council documents' (96/705/EC, ECSC, EURATOM), Official Journal L-325/19, http://eur-lex.europa.eu/LexUriServ/LexUriServ.do?uri=CELEX:31996D0705:EN:HTML (25 May 2012).
Council of the European Union (2000a) 'Report by the Presidency and the General Secretariat of the Council to the European Council on making the co-decision procedure more effective' (13316/1/00 REV 1, LIMITE, CODEC 875), 28 November 2000, www.consilium.europa.eu/uedocs/cms_data/docs/pressdata/en/misc/13316-r1.en0.html (25 May 2012).
Council of the European Union (2000b) 'Council Decision of 6 December 1999 on the improvement of information on the Council's legislative activities and the public register of Council documents' (2000/23/EC), Official Journal L-009/22, 13 January 2000, http://eur-lex.europa.eu/LexUriServ/LexUriServ.do?uri=OJ:L:2000:009:0022:0023:EN:PDF (25 May 2012).
Council of the European Union (2000c) 'Council Decision of 14 August 2000 amending Decision 93/731/EC on public access to Council documents and Council Decision 2000/23/EC on the improvement of information on the Council's legislative activities and the public register of Council documents' (2000/527/EC), Official Journal L-212/9, 23 August 2000, http://eur-lex.europa.eu/LexUriServ/LexUriServ.do?uri=OJ:L:2000:212:0009:0010:EN:PDF (25 May 2012).
Council of the European Union (2000d) 'Proposal for a regulation of the European Parliament and of the Council regarding public access to European Parliament, Council and Commission documents' (14938/00) Presidency Note to delegations, 22 December 2000, www.statewatch.org/secret/14938.PDF (25 May 2012).
Council of the European Union (2001) 'Council Decision of 19 March 2001 adopting the Council's security regulations' (2001/264/EC), Official Journal L-101/1, 11 April 2001, www.eulex-kosovo.eu/training/security/docs/COUNCIL_DECISION_sec.pdf (25 May 2012).
Council of the European Union (2002a) 'Public access to documents – issues of principle. Note from the General Secretariat of the Council to Coreper (Part II)' (6203/02), 1 March 2002, www.statewatch.org/news/2002/mar/6203.pdf (25 May 2012).

Council of the European Union (2002b) 'Seville European Council 21 and 22 June 2002. Presidency Conclusions. Cover note from the Presidency to the delegations' (13463/02), 24 October 2002, www.consilium.europa.eu/ueDocs/cms_Data/docs/pressData/en/ec/72638.pdf (25 May 2012).
Council of the European Union (2004) 'Cover Note from the General Secretariat of the Council to the Working Party on information regarding public access to documents' (11579/04), 2 September 2004, http://register.consilium.europa.eu/pdf/en/04/st11/st11579.en04.pdf (25 May 2012).
Council of the European Union (2005a) 'Council Annual Report on access to documents – 2004' http://consilium.europa.eu/uedocs/cmsUpload/new08896.en05INT.pdf (25 May 2012).
Council of the European Union (2005b) 'Draft Third Annual Report of the Council on the implementation of Regulation No. 1049/2001 of the European Parliament and of the Council of 30 May 2001 Regarding public access to European Parliament, Council and Commission documents' (8896/05), 12 May 2005, http://register.consilium.europa.eu/pdf/en/05/st08/st08896.en05.pdf (25 May 2012).
Council of the European Union (2005c) 'Improving openness and transparency in the Council – draft conclusions. I/A Item note from the Antici Group to Coreper/Council' (15834/05), 15 December 2005, http://register.consilium.eu.int/pdf/en/05/st15/st15834.en05.pdf (25 May 2012).
Council of the European Union (2006a) 'Council Decision of 15 September 2006 adopting the Council's Rules of Procedure' (2006/683/EC, Euratom), Official Journal L-285/47, 16 October 2006, http://eur-lex.europa.eu/LexUriServ/LexUriServ.do?uri=OJ:L:2006:285:0047:0071:EN:PDF (25 May 2012).
Council of the European Union (2006b) 'About the EU – The co-decision procedure', Austria 2006 Presidency of the European Union, www.eu2006.at/en/About_the_EU/Decision_Making/Co_decision_procedure.html (25 May 2012).
Council of the European Union (2006c) 'Council Decision of 17 July 2006 amending Decision 1999/468/EC laying down the procedures for the exercise of implementing powers conferred on the Commission' (2006/512/EC), Official Journal L-200/11, 22 July 2006, http://eur-lex.europa.eu/LexUriServ/LexUriServ.do?uri=OJ:L:2006:200:0011:0013:EN:PDF (25 May 2012).
Council of the European Union (2006d) 'Cover note from the Presidency to the delegations on Brussels European Council 15–16 June 2006, Presidency Conclusions' (10633/06), 17 July 2006, www.consilium.europa.eu/ueDocs/cms_Data/docs/pressData/en/ec/90111.pdf (25 May 2012).
Council of the European Union (2006e) 'An overall policy on transparency. Information sheet from the European Council 15–16 June 2006', 16 June 2006, www.consilium.europa.eu/ueDocs/cms_Data/docs/pressData/en/misc/90112.pdf (25 May 2012).
Council of the European Union (2008) 'Provisional agenda for the 2852nd meeting of the Council of the European Union "Competitiveness" (internal market, industry and research)', 21 February 2008, http://register.consilium.europa.eu/pdf/en/08/st06/st06621.en08.pdf (25 May 2012).
Council of the European Union (2009) 'Council Decision of 1 December 2009 adopting the Council's Rules of Procedure' (2009/937/EU), Official Journal L-325/35, 11 December 2009, http://eur-lex.europa.eu/LexUriServ/LexUriServ.do?uri=OJ:L:2009:325:0035:0061:en:PDF (25 May 2012).
Council of the European Union (2012) 'Outcomes of proceedings. From: Secretariat. To: Delegations. Subject: proposal for a regulation of the European Parliament and of the

Council regarding public access to European Parliament, Council and Commission documents – Meetings on 9 and 27 March 2012of the Working Party on Information' 8410/12, 30 March 2012, www.statewatch.org/news/2012/mar/coe-reg-access-docs.pdf (25 May 2012).

Council internet sites

Council (I), Council register of documents, 'Agendas of Meetings', http://register.consilium.europa.eu/servlet/driver?page=Result&typ=Agendas&lang=EN&fc=REGAISEN&srm=50&md=400&ssf=DATE_DOCUMENT+DESC&ii_PUBLIC_DOC=%3E0&ff_COTE_MATIERE_PRIM=OJ/CONS&cmsid=644 (25 May 2012).

Council (II), Council Register of documents, 'Council Minutes', www.consilium.europa.eu/documents/legislative-transparency/council-minutes?lang=en (25 May 2012).

Council (III), Council Register of documents, Public votes, 'Voting charts', http://register.consilium.europa.eu/servlet/driver?page=Result&typ=Votes&ii_PUBLIC_DOC=%3E0&srm=25&md=400&ssf=FT_DATE+DESC&lang=EN&fc=REGAISEN&ff_SOUS_COTE_MATIERE=vote&cmsid=1323&ff_COTE_DOCUMENT=*%2F12 (25 May 2012).

Council (IV), Council Register of documents 'Monthly summaries of Council acts', www.consilium.europa.eu/documents/legislative-transparency/monthly-summaries-of-council-acts?lang=en (25 May 2012).

Council (V), Council Register of documents, 'Press releases', www.consilium.europa.eu/press/press-releases/latest-press-releases (25 May 2012).

Council (VI), Council Register of documents, www.consilium.europa.eu/documents/access-to-council-documents-public-register?lang=en (25 May 2012).

Council (VII), Council co-decision state by stage website, www.consilium.europa.eu/policies/ordinary-legislative-procedure/search-in-database?lang=en (25 May 2012).

Council (VIII), 'Guide to the ordinary legislative procedure', www.consilium.europa.eu/uedocs/cmsUpload/QC3109179ENC.pdf (25 May 2012).

European Commission documents

European Commission (1994) 'Commission Decision on 8 February 1994 on public access to Commission documents' (94/90/ECSC, EC, Euratom), Official Journal L-46/58, 18 February1994, http://eur-lex.europa.eu/LexUriServ/LexUriServ.do?uri=CELEX:31994D0090:EN:HTML (25 May 2012).

European Commission (2000) 'Proposal for a regulation of the European Parliament and the Council regarding public access to European Parliament, Council and Commission documents' (COM(2000)30 final/2), 21 February 2000, http://eur-lex.europa.eu/LexUriServ/site/en/com/2000/com2000_0030en02.pdf (25 May 2012).

European Commission (2001) 'European governance. A White Paper' (COM (2001) 428), 25 July 2001, http://eur-lex.europa.eu/LexUriServ/site/en/com/2001/com2001_0428en01.pdf (25 May 2012).

European Commission (2002) 'Communication from the Commission. Towards a reinforced culture of consultation and dialogue –general principles and minimum standards for consultation of interested parties by the Commission' (COM(2002) 704 final), 11 December 2002, www.cipra.org/alpknowhow/publications/ciprapublication.2006–01–03.1424172335/Q5com020704en01 (25 May 2012).

European Commission (2005) 'Commission decision of 15 November 2005 amending its Rules of Procedure' (2005/960/EC, Euratom), Official Journal L-347/83, 30 December 2005, http://eur-lex.europa.eu/LexUriServ/site/en/oj/2005/l_347/l_34720051230en00830090.pdf (25 May 2012).
European Commission (2006), 'General report on the activities of the European Union 2005', http://europa.eu/generalreport/pdf/rg2005_en.pdf (25 May 2012).
European Commission (2007) 'General report on the activities of the European Union 2006', http://europa.eu/generalreport/pdf/rg2006_en.pdf (25 May 2012).
European Commission (2008a) 'General report on the activities of the European Union 2007', http://europa.eu/generalreport/pdf/rg2007_en.pdf (25 May 2012).
European Commission (2008b) 'Proposal for a regulation of the European Parliament and of the Council regarding public access to European Parliament, Council and Commission documents' (COM(2008)229final), 30 April 2008, http://eur-lex.europa.eu/LexUriServ/LexUriServ.do?uri=COM:2008:0229:FIN:EN:PDF (25 May 2012).
European Commission (2009) 'General report on the activities of the European Union 2008', http://europa.eu/generalreport/pdf/rg2008_en.pdf (25 May 2012).
European Commission (2010) 'General report on the activities of the European Union 2009', http://europa.eu/generalreport/pdf/rg2009_en.pdf (25 May 2012).
European Commission (2011) 'General report on the activities of the European Union 2010', http://europa.eu/generalreport/en/2010/index.html (25 May 2012).

European Commission internet sites

European Commission (I), Legal bases OLP and special legislative procedures, http://ec.europa.eu/codecision/docs/legal_bases_en.pdf (25 May 2012).
European Commission (II), 'Ordinary Legislative Procedure step by step', http://ec.europa.eu/codecision/stepbystep/text/index_en.htm (25 May 2012).
European Commission (III), 'Your voice in Europe', http://ec.europa.eu/yourvoice/index_en.htm (25 May 2012).
European Commission (IV), 'Prelex: monitoring of the decision-making process between institutions', http://ec.europa.eu/prelex/apcnet.cfm?CL=en (25 May 2012).
European Commission (V), 'Register of Commission documents',
http://ec.europa.eu/transparency/regdoc/registre.cfm?CL=en (25 May 2012).
European Commission (VI), 'Co-decision or the Ordinary Legislative Procedure', http://ec.europa.eu/codecision/index_en.htm (25 May 2012).
European Commission (VII), 'Help on document type' http://ec.europa.eu/transparency/regdoc/aidetypesdoc.cfm?CL=en (25 May 2012).
European Commission (VIII), 'Codecision step by step', http://ec.europa.eu/codecision/stepbystep/index_en.htm (25 May 2012).

European Council

European Council (2002) 'Presidency conclusions, Seville European Council, 21 to 22 June 2002' (13463/02), 24 October 2002, http://ue.eu.int/ueDocs/cms_Data/docs/pressData/en/ec/72638.pdf (25 May 2012).

European Ombudsman documents

European Ombudsman (2005a) 'Draft recommendation to the Council of the European Union in complaint' (1487/2005/GG), 14 March 2006, www.ombudsman.europa.eu/recommen/en/051487.htm (25 May 2012).

European Ombudsman (2005b) 'Special Report from the European Ombudsman to the European Parliament following the draft recommendation to the Council of the European Union in complaint' (2395/2003/GG), 04 October 2005, www.ombudsman.europa.eu/cases/specialreport.faces/en/386/html.bookmark (25 May 2012).

European Ombudsman internet sites

European Ombudsman (I), 'Could he help you?', www.ombudsman.europa.eu/media/en/default.htm#html (25 May 2012).

European Ombudsman (II), "Complaint form" https://secure.ombudsman.europa.eu/atyourservice/secured/complaintform/home.faces (25 May 2012).

European Parliament documents

European Parliament (1994) 'Decision of the European Parliament of 9 March 1994 on the regulations and general conditions governing the performance of the Ombudsman's duties' (94/262), Official Journal L-113/15, 4 May 1994, http://eur-lex.europa.eu/LexUriServ/LexUriServ.do?uri=CELEX:31994D0262:EN:HTML (25 May 2012).

European Parliament (2002) 'Joint conference on co-decision and conciliation, Monday 4 and Tuesday 5 November 2002 Brussels, European Parliament. Chaired by Giorgos Dimitrakopoulos, Charlotte Cederschiöld and Renzo Imbeni, Vice-Presidents of the European Parliament. Informal record of the discussions', www.europarl.europa.eu/code/events/20021104/minutes_en.pdf (25 May 2012).

European Parliament (2004) 'Activity report 1 May 1999 to 30 April 2004 (5th parliamentary term) of the delegations to the Conciliation Committee presented by Giorgos Dimitrakopoulos, Charlotte Cederschiöld and Renzo Imbeni, Vice Presidents' (DV\530227EN), www.europarl.europa.eu/code/information/activity_reports/activity_report_1999_2004_en.pdf (25 May 2012).

European Parliament (2004) 'Conciliation and co-decision. A guide to how Parliament co-legislates' (DV/547830EN), November 2004, on file with the author.

European Parliament (2006) 'Report with recommendations to the Commission on access to the institutions' texts' (FINAL A6–0052/2006), Committee on Civil Liberties, Justice and Home Affairs, www.europarl.europa.eu/omk/sipade3?PUBREF=-//EP//NONSGML+REPORT+A6-2006-0052+0+DOC+PDF+V0//EN&L=EN&LEVEL=2&NAV=S&LSTDOC=Y (25 May 2012).

European Parliament (2007a) 'Conciliations and co-decision activity report. July 2004 to December 2006 (6th parliamentary term. First half-term)' (DV/651053), 31 January 2007, www.europarl.europa.eu/code/information/activity_reports/activity_report_2004_2006_en.pdf (25 May 2012).

European Parliament (2007b) 'Conciliations and co-decision. A Guide to how the Parliament co-legislates' (DV/684001EN), July 2007, on file with the author.

European Parliament (2009a) 'Co-decision and conciliation. A Guide to how the Parliament co-legislates under the Treaty of Lisbon' (DV\796687EN) November 2009, on file with the author.

European Parliament (2009b) 'Activity report 1 May 2004 to 13 July 2009 (6th parliamentary term) of the delegations to the Concililation Committee presented by Rodi Kratsa-Tsagaropoulou, Alejo Vidal-Quadras and Mechtild Rothe. Vice-Presidents responsible for conciliation' www.europarl.europa.eu/code/information/activity_reports/activity_report_2004_2009_en.pdf (25 May 2012).

European Parliament (2009c) 'Proposal for a regulation of the European Parliament and of the Council regarding public access to European Parliament, Council and Commission documents (recast) (COM(2008)0229 – C6–0184/2008 – 2008/0090(COD))', A6–0077/2009, 29 February 2009, www.europarl.europa.eu/sides/getDoc.do?pubRef=-//EP//NONSGML+REPORT+A6–2009–0077+0+DOC+PDF+V0//EN (25 May 2012).

European Parliament (2011) 'Annual report 2010 on public access to documents (Regulation (EC) No 1049/2001 – Article 17)', www.europarl.europa.eu/RegData/PDF/862177_EN.pdf (25 May 2012).

European Parliament (2012) 'Rules of procedure for the European Parliament' seventh parliamentary term – May 2012, www.europarl.europa.eu/sides/getDoc.do?pubRef=-//EP//NONSGML+RULES-EP+20120521+0+DOC+PDF+V0//EN&language=EN.

Committee on Civil Liberties, Justice and Home Affairs PE415.164.

European Parliament internet sites

EP (I), European Parliament register of documents, www.europarl.europa.eu/RegistreWeb/search/typedoc.htm?language=EN (25 May 2012).

EP (II), European Parliament conciliations and co-decision homepage "Conciliations and Codecision homepage Ongoing Conciliations" www.europarl.europa.eu/code/conciliation/default_en.htm (25 May 2012).

EP (III), European Parliament conciliations and co-decision homepage "Conciliation and Codecision", www.europarl.europa.eu/code/default_en.htm (25 May 2012).

EP (IV), 'The legislative observatory (OEIL)', www.europarl.europa.eu/oeil/home/home.do (25 May 2012).

EP (V), 'European Parliament register of documents', www.europarl.europa.eu/RegWeb/application/registre/simpleSearch.faces?language=EN (25 May 2012).

EP (VI), European Parliament conciliations and co-decision homepage, 'Ongoing procedures' www.europarl.europa.eu/code/ongoing/default_en.htm (25 May 2012).

EP (VII), European Parliament 'Meeting documents of the Committee on Economic and Monetary Affairs' www.europarl.europa.eu/meetdocs/2009_2014/organes/econ/econ_7leg_meetinglist.htm (25 May 2012).

EP (VIII), European Parliament 'ENVI-News – newsletter from the European Parliament Committee on Environment, Consumers and Public Health Committee' www.europarl.europa.eu/committees/en/envi/newsletters.html (25 May 2012).

EP (IX), European Parliament 'RegiNews – Newsletter from the European Parliament Committee on Regional Development' www.europarl.europa.eu/committees/en/regi/newsletters.html (25 May 2012).

EP (X) European Parliament, Conciliations and Co-decision homepage "Codecison news" www.europarl.europa.eu/code/newsletter/default_en.htm(25 May 2012).

Inter-institutional documents

European Parliament and Council of the European Union (2001) 'Regulation 1049/2001 of the European Parliament and the Council regarding public access to European Parliament, Council and Commission documents', Official Journal L-145/43, 31 May 2001, www.consilium.europa.eu/uedocs/cmsUpload/L14520EN.pdf (25 May 2012).

European Parliament and Council of the European Union (2006a) 'Regulation 562/2006 establishing a Community Code on the rules governing the movement of persons across borders (Schengen Borders Code)', Official Journal L-105/1, 13 April 2006, http://eur-lex.europa.eu/LexUriServ/LexUriServ.do?uri=OJ:L:2006:105:0001:0032:EN:PDF (25 May 2012).

European Parliament and Council of the European Union (2006b) 'Regulation 1987/2006 of the European Parliament and of the Council of 20 December 2006 on the establishment, operation and use of the second generation Schengen Information System (SIS II)', Official Journal L-381/4, 28 December 2006, http://eur-lex.europa.eu/LexUriServ/LexUriServ.do?uri=OJ:L:2006:381:0004:0023:EN:PDF(25 May 2012).

European Parliament, Council and Commission (1999) 'Joint declaration on practical arrangements for the new co-decision procedure (Article 251 of the Treaty Establishing the European Community)', Official Journal C-148/1, 28 May1999, http://eur-lex.europa.eu/LexUriServ/site/en/oj/1999/c_148/c_14819990528en00010002.pdf (25 May 2012).

European Parliament, Council, Commission (2007), 'Joint declaration on practical arrangements for the co-decision procedure (article 251 of the EC Treaty)', (2007/C 145/02), Official Journal C-145/6, 30 June 2007, http://eur-lex.europa.eu/LexUriServ/LexUriServ.do?uri=OJ:C:2007:145:0005:0009:EN:PDF (25 May 2012).

European Union (2004) 'Treaty establishing a Constitution for Europe', Official Journal C-310/01, 16 December 2004, http://eur-lex.europa.eu/JOHtml.do?uri=OJ:C:2004:310:SOM:EN:HTML(35 May 2012).

European Union (2006) 'Consolidated versions of the Treaty on European Union and of the Treaty establishing the European Community', Official Journal C-321, 29 December 2006, http://eur-lex.europa.eu/LexUriServ/LexUriServ.do?uri=OJ:C:2006:321E:0001:0331:EN:pdf (25 May 2012).

European Union (2010) 'Consolidated versions of the Treaty on European Union and the Treaty on the Functioning of the European Union' Official Journal C-83/1, 30 March 2010, http://eur-lex.europa.eu/LexUriServ/LexUriServ.do?uri=OJ:C:2010:083:FULL:EN:PDF (25 May 2012).

Other EU/inter-institutional internet sites

Europa Glossary (I), 'Codecision Procedure', http://europa.eu/legislation_summaries/glossary/codecision_procedure_en.htm (25 May 2012).

EUR-Lex service, 'Access to European Union Law', http://eur-lex.europa.eu/en/index.htm (25 May 2012).

Other documents

Public sphere

Access Info (2012), see website on 'Renewed threat on access to EU documents – EU Transparency Campaign' www.access-info.org/en/european-union/234-renewed-threat-on-access-to-eu-documents (25 May 2012).

EurActive (2006) 'Stakeholder consultation: A voice for civil society in Europe?' published 16 June and updated 26 June 2006, www.euractiv.com/en/pa/stakeholder-consultation-voice-civil-society-europe/article-156178 (25 May 2012).

EurActive (2006) 'Transparency initiative' published 8 June 2005, updated 20 September 2006, www.euractiv.com/en/pa/transparency-initiative/article-140650 (25 May 2012).

EurActive (2006) 'MEPs urge EU Council to prop up transparency' published 26 February 2006, www.euractiv.com/priorities/meps-urge-eu-council-prop-transparency/article-152940 (25 May 2012).

EurActive (2006) 'Finnish Presidency to shed light on Council workings' published 5 April 2006, www.euractiv.com/priorities/finnish-presidency-shed-light-council-workings/article-153906 (25 May 2012).

European Voice (2001) 'Openness is still a closed case in some institutions' 10 May 2001, www.europeanvoice.com/article/imported/openness-is-still-a-closed-case-in-some-institutions/42692.aspx (25 May 2012).

European Voice (2001) 'Anti-secrecy MEPs snub transparency' 12 July 2001, www.europeanvoice.com/article/imported/anti-secrecy-meps-snub-transparency/43065.aspx (25 May 2012).

European Voice (2001) 'Council backs down after four-year fight over access to papers' 26 July 2001, www.europeanvoice.com/article/imported/council-backs-down-after-four-year-fight-over-access-to-papers/43155.aspx (25 May 2012).

European Voice (2002) 'German and Danish MEPs in U-turn over online declaration' 17 January 2002, www.europeanvoice.com/article/imported/german-and-danish-meps-in-u-turn-over-online-declaration/43969.aspx (25 May 2012).

European Voice (2004) 'Poor visitor access "undermines Parliament transparency"' 1 April 2004, www.europeanvoice.com/article/imported/popularity-of-parliament-is-cause-for-pride/49666.aspx (25 May 2012).

European Voice (2005) 'Child medicine plan misses mark' 21 April 2005, www.europeanvoice.com/article/imported/child-medicine-plan-misses-mark/52182.aspx (25 May 2012).

European Voice (2007) 'Consultancies spurn lobbying register' 10 May 2007, www.europeanvoice.com/article/imported/consultancies-spurn-lobbying-register/57402.aspx (25 May 2012).

Statewatch (2001a) 'Open letter from civil society on the new code of access to documents of the EU institutions' 2 May 2001, www.statewatch.org/news/2001/apr/13openletter.htm (25 May 2012).

Statewatch (2001b) 'Statewatch wins new complaint against the Council' www.statewatch.org/news/2001/jul/ombuds.htm (25 May 2012).

Statewatch (2002) 'US government vetoes Statewatch request for access to EU-US agendas' www.statewatch.org/news/2002/jan/03usveto.htm (25 May 2012).

Statewatch (2002) 'Council of the European Union disagrees on giving access to the public of positions taken by EU governments' www.statewatch.org/news/2002/mar/02open.htm (25 May 2012).

Statewatch (2003) 'EU annual reports on access to documents – still a very long way to go', speech by Tony Bunyan to the European Parliament 12 June 2003, www.statewatch.org/news/2003/jun/24open.htm (25 May 2012).

Statewatch (2005) 'More openness or just a drop in the ocean? – the need for freedom of information in the EU' Viewpoint by Tony Bunyan, www.statewatch.org/news/2005/dec/03foi-in-eu.htm (25 May 2012).

Statewatch Observatory on the Regulation on access to documents: 2008-ongoing, www.statewatch.org/foi/observatory-access-reg-2008–2009.htm (25 May 2012).

Index

Page numbers in *italics* denote tables, those in **bold** denote figures.